"Dr. Jason Tham's work with design thinking provides a model for all voices to be heard within an innovative problem-solving framework that connects rigorous research with the spirit of an entrepreneurial mindset. A compelling read for anyone considering the future of technical communication and the ways in which the field can embrace radical collaborations to untangle wicked social problems."
— **Dr. Laura A. Palmer**, *Chair and Professor of Technical Communication and Rhetoric, Kennesaw State University*

"*Design Thinking in Technical Communication: Solving Problems Through Making and Collaboration* is a book that traverses the dual genres of theory and praxis. What Tham does in this book is significant for our field: he identifies design thinking as a way to address the social aspect of technical communication, which is rapidly becoming a touchstone for addressing contemporary problems. Viewing technical communication as a catalyst for invention and innovation allows us to introduce students to ways they can solve complex problems through design thinking and making. What I appreciate about this text is that Tham addresses both the theoretical and practical aspects of his topic. He provides pedagogical strategies at the end of each chapter, which demonstrate how practitioners can enact the ideas he presents in their own classrooms. Moreover, the focus on empathy and radical collaboration as components of technical communication practice reinforces the need to reframe our field in ways that focus on the human condition. This book serves as a framework for seeing our field as a site for positive changemaking. I would recommend this text as a good introduction to the concept of design thinking, as well as ways to integrate it into classrooms."
— **Dr. Jennifer Bay**, *Associate Professor of English and Director of Professional Writing, Purdue University*

DESIGN THINKING IN TECHNICAL COMMUNICATION

This book explicates the relationships between design thinking, critical making, and socially responsive technical communication. It leverages the recent technology-powered DIY culture called the "Maker Movement" to identify how citizen innovation can inform cutting-edge social innovation that advocates for equitable change and progress on today's "wicked" problems.

After offering a succinct account of the origin and recent history of design thinking, along with its connections to the design paradigm in writing studies, the book analyzes maker culture and its influences on innovation and education through an ethnographic study of three academic makerspaces. It offers opportunities to cultivate a sense of critical changemaking in technical communication students and practitioners, showcasing examples of socially responsive innovation and expert interviews that urge a disciplinary attention to social justice advocacy and an embrace of the design thinking principle of radical collaboration. The value of design thinking methodologies for teaching and practicing socially responsible technical communication are demonstrated as the author argues for a future in the field that sees its constituents as leaders in radical innovation to solve wicked social problems.

This book is essential reading for instructors, students, and practitioners of technical communication, and can be used as a supplemental text for graduate and undergraduate courses in usability and user-centered design and research.

Jason Chew Kit Tham (PhD, University of Minnesota) is an assistant professor of technical communication and rhetoric in the Department of English and co-director of the User Experience Research Lab at Texas Tech University. He teaches courses in user experience research, information design, and digital rhetoric. He is a winner of the 2020 CCCC Emergent Researcher Award and the 2019 STC Frank R. Smith Award for Best Article in *Technical Communication*. His research has been published in journals such as *Technical Communication Quarterly, Journal of Business and Technical Communication, Journal of Technical Writing and Communication, IEEE Transactions on Professional Communication, Communication Design Quarterly*, and *Computers and Composition*. He is the co-author of *Collaborative Writing Playbook: An Instructor's Guide to Designing Writing Projects for Student Teams* (Parlor Press).

ATTW Book Series in Technical and Professional Communication

Tharon Howard, Series Editor

Creating Intelligent Content with Lightweight DITA
Carlos Evia

Editing in the Modern Classroom
Michael J. Albers and Suzan Flanagan

Translation and Localization: A Guide for Technical Communicators
Bruce Maylath, and Kirk St.Amant

Technical Communication After the Social Justice Turn: Building Coalitions for Action
Rebecca Walton, Kristen R. Moore and Natasha N. Jones

Content Strategy in Technical Communication
Guiseppe Getto, Jack T. Labriola and Sheryl Ruszkiewicz

Teaching Content Management in Technical and Professional Communication
Tracy Bridgeford

Design Thinking in Technical Communication
Solving Problems through Making and Collaboration
Jason Chew Kit Tham

For additional information on this series please visit www.routledge.com/ATTW-Series-in-Technical-and-Professional-Communication/book-series/ATTW, and for information on other Routledge titles visit www.routledge.com.

DESIGN THINKING IN TECHNICAL COMMUNICATION

Solving Problems through Making and Collaboration

Jason Chew Kit Tham

Routledge
Taylor & Francis Group

NEW YORK AND LONDON

First published 2021
by Routledge
52 Vanderbilt Avenue, New York, NY 10017

and by Routledge
2 Park Square, Milton Park, Abingdon, Oxon, OX14 4RN

Routledge is an imprint of the Taylor & Francis Group, an informa business

© 2021 Taylor & Francis

Library of Congress Cataloging-in-Publication Data
Names: Tham, Jason Chew Kit, 1990– author.
Title: Design thinking in technical communication : solving problems through making and collaboration / Jason Chew Kit Tham.
Description: New York, NY : Routledge, 2021. | Series: ATTW series in technical and professional communication | Includes bibliographical references and index.
Identifiers: LCCN 2020054734 (print) | LCCN 2020054735 (ebook) | ISBN 9780367478322 (hardback) | ISBN 9780367478216 (paperback) | ISBN 9781003036760 (ebook)
Subjects: LCSH: Communication of technical information. | Industrial design. | Makerspaces. | Intercultural communication.
Classification: LCC T10.5 .T465 2021 (print) | LCC T10.5 (ebook) | DDC 601/.4—dc23
LC record available at https://lccn.loc.gov/2020054734
LC ebook record available at https://lccn.loc.gov/2020054735

ISBN: 978-0-367-47832-2 (hbk)
ISBN: 978-0-367-47821-6 (pbk)
ISBN: 978-1-003-03676-0 (ebk)

Typeset in Bembo
by Apex CoVantage, LLC

To Ann Hill Duin—thank you for teaching me about being a scholar and teacher, and for trusting me with your collaboration.

CONTENTS

List of Figures   *xi*
List of Tables   *xii*
Preface   *xiii*
Acknowledgements   *xvi*
Series Editor Foreword to Design Thinking in Technical
Communication   *xviii*

1   Introducing Design Thinking (and Making) for Technical
    Communication   1

2   The Maker Movement and Its Influences on Technical
    Communication and Higher Learning: A Look at Three
    Makerspaces   27

3   Social Innovation: Designing Humane Technical
    Communication   57

4   Making and Design Thinking as Pedagogical Strategies
    for Social Advocacy   79

5   Cultivating Radical Collaboration in Technical
    Communication   99

Conclusion: Disrupting and Innovating in Technical
Communication Through Making and Design Thinking

121

Appendix A: Radical Collaboration Survey Questionnaire   *131*
Appendix B: Design Thinking Methods and Exercises   *133*
Appendix C: Annotated Bibliography   *137*
Index   *144*

FIGURES

1.1 Key developments of design thinking   6
1.2 The basic model of design thinking adapted from the Stanford
     model (Stanford d.school, n.d.)   10
2.1 A simple makerspace setup in a classroom   53
3.1 A tweet by Vivianne Castillo on June 23, 2019 asking UX
     professionals to talk about privilege, racism, homophobia,
     white supremacy, xenophobia, etc.   60
3.2 Design thinking provides a methodology for socially just and
     ethical innovation   71
3.3 A simple empathy map for capturing human experience   73
4.1 The assignment sequence in WRIT 3562W mapped onto the
     design thinking process   83
5.1 A screenshot of the WRC website in 2015   104
5.2 The iterative process of collaborative autoethnography in this study   106
5.3 Comparison of degree of actualization (mean and median)
     across six radical collaboration dimensions   112
5.4 A model for radical collaboratory across programs, institutions,
     and disciplines   113

TABLES

4.1   The design thinking process and correlating assignments and
goals in each phase   84
4.2   Major assignment descriptions and weight in percentage   85
4.3   Overview of student design challenge projects and outcomes   91
5.1   Participants' academic rank at the time of survey   107
5.2   Participants' responses to Question 1   108
5.3   Degree of actualization of radical collaboration dimensions   112
5.4   Strategies for cultivating radical collaboration across disciplinary,
institutional, and programmatic contexts   114
5.5   Radical collaboration auditing worksheet   117

PREFACE

Like many of my colleagues in technical communication and rhetoric, I have always struggled to tell my family and friends outside the academic world (and some inside) what it is that I do and teach. Usually, I would say that I teach communication, focusing more on the artifacts in the communication process rather than human relations. But that is not at all accurate. In fact, my work is all about being user centered and audience aware. I pay attention to human psychology and how people interact with one another. A majority of my research focuses on the process rather than the product of communication. So, I began telling people that I teach design instead (since most of my course titles have "design" in it), and that I am a designer of communication.

That didn't exactly make it easier for others to understand my work. I now face questions like, "So, you teach graphic design?" "Do you build websites?" "Can you help me with a logo?" etc. I realize the term "design" is just as open-ended as "communication." So is the subject matter of this book, "design thinking." Common misconceptions about design thinking include:

- Design thinking is visual thinking.
- Design thinking is about paying attention to the look of a product.
- Design thinking is all about the usability of a product.
- Design thinking is a protocol for product development.
- Design thinking is about making people more creative.
- Design thinking is thinking about design (well, this is not entirely wrong).

Believe it or not, these are comments I have gathered from folks who teach and practice technical communication. Having completed a dissertation that employed design thinking, I recognized an opportunity to write a book about design thinking that hopefully helps to mitigate some of these misconceptions.

Design Thinking in Technical Communication modernizes technical communication pedagogy and practice by highlighting the connections and overlaps between design thinking, making, and technical communication. It takes advantage of the recently popularized "Maker Movement" to forward the argument that novice invention can facilitate social innovation through user-centered design and collaboration. This book examines the underlying design thinking principles in maker culture that enable innovative problem solving while offering opportunities to cultivate empathy in technical communication students and practitioners. The investigations featured in this book reveal the values of design thinking and making for teaching and practicing user-centered technical communication. Moreover, I argue for a future of technical communication that sees its constituents as leaders in productive disruption to solve complex problems. This book provides pedagogical as well as practical strategies for achieving this goal.

Intended Audience

I envisioned the primary readers of this book to be instructors and students of undergraduate and graduate-level technical communication courses (including proseminars on user-centered design, user experience research, and usability studies). Instructors can use the pedagogical materials and strategies in this book to design and deliver design thinking assignments that align with technical communication objectives. Undergraduate students may use this book as a special introduction to technical communication through the lens of design. Graduate students in technical communication and rhetoric can use this book as a resource for their seminar papers, comprehensive exams, and dissertation. For career-oriented students who are studying user research and user experience design, this book will make a timely complement to titles in usability and user-centered communication design.

Industry leaders and technical communication practitioners are the secondary audience of this book. These professionals may learn from the case studies and discussions in this book how the academy is connecting theory and practice, and vice versa, in terms of design, making, and innovation. Practitioners can expect to learn how design thinking fits within the greater agenda of the field of technical communication.

Overview of Chapters

Each chapter of this book scaffolds an aspect of design thinking and its material manifestation, "making," by situating them in existing literature and drawing new insights from recent studies.

Chapter 1 introduces and defines design thinking, and traces its relationship to technical communication through the design paradigm in writing studies. It includes a brief account on the origin of design thinking and a description of its methodology and mindset. Through the lens of user-centered design, I establish an initial connection between design thinking and making as a problem-solving strategy for technical communication pedagogy and practice.

Chapter 2 provides an in-depth discussion on the development of making as a cultural movement and its influence on novice innovation and higher education today. Through an ethnographic study of three academic makerspaces, I compare three prominent campus makerspaces and their respective operational strategies. To show the benefits of these spaces, I present multiple student accounts on the impact of making on peer-to-peer collaboration, mentoring, and community engagement.

Chapter 3 forwards the argument that design thinking can help technical communicators to be stronger user advocates through social innovation. By showcasing examples of socially responsive innovation and a collection of industry expert perspectives, I urge scholars and industry practitioners to establish a disciplinary agenda to lead social justice advocacy efforts.

Chapter 4 demonstrates how design thinking and making can cultivate social innovation from within the technical communication classroom. It features a pedagogical case study that details the process in which students addressed a community problem and created tangible solutions using the design thinking methodology.

Chapter 5 zeroes in on a specific attribute of design thinking, i.e., radical collaboration, and offers insights from a self-study of a research collaboratory. It reveals the challenges and strategies in cultivating authentic collaboration through the tenets of radical collaboration. To support this culture of collaboration, I provide a sample three-tier model focusing on the academic condition.

The concluding chapter invites scholars and practitioners to participate in making, disrupting, and innovating change through design thinking. I argue that making and design thinking provide productive frameworks for positive change-making, and that technical communication is well-positioned to pursue this direction.

This book is supplemented with two pedagogical appendices. The first appendix contains some signature design exercises and research methods employed by design thinkers and makers. The second appendix is an annotated bibliography of 15 key readings mentioned in the book and their respective annotated summaries.

ACKNOWLEDGEMENTS

This book project has benefited from the support and feedback of many smart, kind, and generous individuals. I would like to use this space to recognize their contributions.

First, I thank Tharon Howard, editor of the Routledge ATTW Book Series in Technical and Professional Communication, for believing in this project and helping me shape it into its current form. I also thank the anonymous reviewers of the manuscript proposal for offering excellent feedback that improved the focus of this work. A heartfelt thank you to Grant Schatzman, editorial assistant at Routledge/Taylor & Francis, and his team for guiding me through the publication process.

To all the students and industry professionals who participated in the studies of this book, thank you for sharing your experiences and insights with me. I thank the makerspace managers at the University of Minnesota Anderson Labs, Georgia Tech Invention Studio, and Case Western Think[box] who hosted and showed me around when I was observing the sites. Thank you to Josh Halverson for introducing me to the key personnel. A special shout out to the ever-so-patient staff members at the University of Minnesota Liberal Arts Technologies and Innovation Services (LATIS), particularly Samantha Porter, for assisting me with multiple deployments of maker technologies in my classes.

I especially thank all my WRIT 3562 Technical and Professional Writing students in the Fall 2017 semester for embracing and engaging with the design challenge study. You inspire me with your creativity and dedication to learning.

Next, I want to thank a number of colleagues and friends who have read and provided relevant feedback on this manuscript. Scott Weedon, Beau Pihlaja, Antonio Byrd, Rob Grace, and Joe Moses—your comments and suggestions helped make this a better book. To my peers with whom I have shared different parts of this book during my research process—bonus points if you were a part of the Emerging Technologies Research Collaboratory—thank you for your listening and vote of confidence when I needed them most.

I am grateful to have been a part of the University of Minnesota's Rhetoric and Scientific and Technical Communication doctoral program, which challenged me intellectually and shaped my scholarly identity. To my amazing advisor Ann Hill Duin, thank you for sharing your expertise and energy with me even after I have graduated from the program. Many thanks to the Writing Studies faculty, including Lee-Ann Kastman Breuch, Laura Gurak, Donald Ross, Tom Reynolds, John Logie, Pat Bruch, Richard Graff, and Christina Haas, for making technical communication such an exciting field of inquiry.

The design thinking community in technical communication has continued to grow. I am thankful for the productive exchanges I got to have with so many smart humans at conferences, over Twitter, and on other occasions about design thinking. These scholars include Emma Rose, Rebecca Walton, Natasha Jones, Kristen Moore, Gustav Verhulsdonck, Isabel Pedersen, Debby Andrews, Liz Lane, Rebecca Pope-Ruark, Kirk St.Amant, Joyce Locke Carter, Cana Uluak Itchuaqiyaq, Adam Strantz, Nadya Shalamova, Tammy Rice-Bailey, Chen Chen, Jack Labriola, Luke Thominet, Halcyon Lawrence, Daniel Richards, Lauren Garskie, Beth Shirley, Lauren Cagle, Josephine Walwema, John Spartz, Maria Novotny, Les Hutchinson, Estee Beck, Andrew Kulak, Laura Gonzales, April O'Brien, Ashanka Kumari, Shannon Butts, Nupoor Ranade, Jason Luther, Jennifer Sano-Franchini, Matthew Vetter, Devon Cook, Jason Markins, Jialei Jiang, Derek Ross, Quan Zhou, and Xiaoli Li.

I thank my mentors and colleagues in the Department of English and the Technical Communication and Rhetoric program at Texas Tech University for their constant encouragement and support for this project.

Finally, I thank my family members, especially my mom for always listening to me talk about my work and life. To my partner and love, Kamm, thank you for cheering me on and giving me time and space to complete this project.

SERIES EDITOR FOREWORD TO *DESIGN THINKING IN TECHNICAL COMMUNICATION*

The ATTW Series in Technical and Professional Communication constantly searches for authors who can blend solid scholarship with pedagogical application on important themes and issues in our field. So, I was thrilled when Dr. Jason Tham approached me about doing a volume for the ATTW Series on Design Thinking and Technical Communication. One reason I was excited is because, as Jason also points out in the book, very little serious, focused attention has been given to the potential impact Design Thinking can have on the field of Technical and Professional Communication—this in spite of the important impact that Design Thinking has had on the practice of designing for users in industry and on the pedagogical training of engineers and graphic designers in programs like the famous d.school at Stanford. Universities like MIT, the Uni. of Minnesota, Elon, Vanderbilt, Georgia Tech, and Case Western (just to name a few) have been making enormous investments in "makerspaces" at their institutions. And yet, while these makerspaces are intended to be open sites where students across campus can create cross-functional teams to find solutions to "wicked" social problems, these resources have largely escaped the examination of technical communicators. Thanks to *Design Thinking in Technical Communication* this is no longer the case. In Chapter Two, Tham takes us on a tour of three of the leading makerspaces at US universities and shows us in practical and pragmatic terms how and why technical communicators really should be participating in the innovation happening there. In short, our field has needed a book-length introduction to Design Thinking and its impact on TPC.

However, another reason that I was excited when Jason approached me about *Design Thinking in Technical Communication* is because I've collaborated with him on journal articles discussing the impact of UX on technical communication practices and I knew that Jason has a deep and thorough grasp of the history and theory of Design Thinking—he is, in my opinion, the best author for the job. One of the complaints long time UX researchers like Don Norman or Jacob Nielsen have offered about Design Thinking proponents is that they often describe Design Thinking as though it was the product of a virgin birth; indeed, one of the most common memes heard from the UX community in response to Design Thinking is "we already do that." Jason's work here is exciting because he eschews the myopia of early introductions of Design Thinking and, in the very first chapter of *Design Thinking in Technical Communication*, he takes us on a masterful tour of the movement's history and theory.

But the third reason I'm excited about *Design Thinking in Technical Communication* is because Tham doesn't stop with the theory and culture which informs Design Thinking, this book melds theory and practice. Every chapter of the book provides practical and pragmatic recommendations for technical communication pedagogy. At the end of Chapter Three, for example, the book offers readers a set of pedagogical exercises in design thinking, including empathy mapping, contextual inquiry, and rapid prototyping, which can be modified or scaled to accommodate different class sizes and degree levels. But ultimately, I think Jason's most important contribution is his return to what usability professionals in the 1990s used to call "user advocacy" and what we now call "social justice." His book invites both students and practitioners to foster change in technical communication projects, to solve social justice (wicked) problems by means of design thinking and making. It returns us to our field's propensity for a human-centered problem-solving orientation and demonstrates how design thinking is already sympathetic to technical communication practices in a way that encourages the technical communication classroom to become what Jason calls sites of "productive disruption."

Again, I'm thrilled that this collection is part of the ATTW Series in Technical and Professional Communication. The topic is totally consistent with and wholly appropriate for the ATTW Series and provides exactly the blend of solid scholarship and pedagogical application that we're seeking in the series.

Tharon W. Howard
Editor, ATTW Series in Technical and Professional Communication
November 3, 2020

1

INTRODUCING DESIGN THINKING (AND MAKING) FOR TECHNICAL COMMUNICATION

Overview: This chapter opens with the argument that design thinking matters now more than ever for technical communication. By way of showcasing the growing prominence of design thinking practices in academia and industry, this introduction presents the basic principles of design thinking and its philosophy for human-centered solutions. It combines key emphases of technical communication—including user experience, user-centered design, usability—with design-centric thinking and the growing tendency for treating technical communicators as problem-solving professionals. The introduction shows the strong overlaps between current technical communication pedagogical approaches and design thinking frameworks. The chapter includes with a preview of the recent novice maker culture that informed what is fondly known as the Maker Movement in education.

The Design (Thinking) Turn

Ongoing narratives in the academic discipline of technical communication urge scholars to pay attention and respond to the evolving nature of technical communication work and practices, tools and technologies, values and cultures (Johnson-Eilola, 1996; Spilka, 2002; Bekins & Williams, 2006; Hailey, Cox, & Loader, 2010; Zhang & Kitalong, 2015). Changing problems in the workplace and the classroom call for innovative thinking and actions, as Linn Bekins and Sean Williams (2006) and Hailey et al. (2010) contended:

> The creative economy has affected technical and professional communication curricula, students, and alumni in ways that have increased the emphasis on technological aptitude, an ability to work with multiple cultures and numerous independent contractors, to deal with changing expectations,

and to manage creative, dynamic, and often nonlinear projects. (Bekins & Williams, 2006, p. 294)

We further suggest that technical communicators who consistently identify and solve important corporate problems and who develop innovations that positively impact the corporation's bottom line will be more valued than those who write well but contribute nothing more.

(Hailey et al., 2010, p. 139)

Indeed, we must prepare students to face unprecedented issues. Recent events including the COVID-19 global health pandemic and ongoing racial tensions due to systemic oppression and xenophobia have proven to technical communication scholars, instructors, and practitioners that the social aspect of technical communication is as important as—in some cases, *more important than*—its technical counterpart. Jason Swarts (2020) argued that technical communicators play the crucial role in socially constructing users' interaction with technologies and with their respective communities. Relatable examples from the COVID-19 crisis are the hand-washing instructions and social distancing guidelines that provided both technical information and social organizing measures to the general public. Evidently, emerging issues in our society, the workplace, and academia, plus the evolution of information technology, have helped technical communication mature into an innovative profession. Our job is no longer just about translating complex technical information for everyday users but instead solving problems through communication and material resources. In this book, I further contend that technical communication serves a catalyst for social innovation and changemaking. This contention aligns with existing technical communication scholarship derived from the larger field of writing studies.

The North American academic tradition of technical communication is heavily influenced by rhetoric and writing studies theories and scholarly developments. Among the most important paradigms in the understanding of contemporary rhetoric and communication were the process turn (circa 1960s), the social and post-process turns (1980s), and the critical and cultural turn (1990s). While these "turns" were not successive but rather staggered, each has borrowed from its predecessor's theoretical bases and assumptions. Fortuitously, the dawn of the social and post-process paradigm was dovetailed by serious discussions among composition specialists and professional communication theorists on the connections between writing and design.[1] Borrowing from the vocabulary of design studies, Charles Kostelnick's (1989) *College Composition and Communication (CCC)* article, "Process Paradigm in Design and Composition: Affinities and Directions," critiqued the then buzzword, "process pedagogy," and offered *design* as a counterpart to the *writing* process. Succinctly, Kostelnick stated,

As a medium for creativity and communication, design is the natural counterpart to writing, one adapting visual, the other verbal, language to diverse contexts. The parallels between the process approaches encompass an array of cross-disciplinary issues central to the creative act. Both process movements have explored creativity as a sequence of interrelated activities and have shifted from linear stage models to recursive cyclic models. Similarly, both movements have treated invention as a problem-solving task adapted to a particular audience and context.

(Kostelnick, 1989, p. 267)

Kostelnick's argument was largely influenced by Richard Buchanan's pioneering efforts in identifying the rhetorical facets of design practices. In "Declaration by Design: Rhetoric, Argument, and Demonstration in Design Practice," Buchanan (1985) forwarded the idea of design as demonstrative rhetoric—that built objects and interfaces "influence and shape society by [their] persuasive assertions" (p. 22). Buchanan considered designers as major contributors to the reconfiguration of new-age rhetoric by "shaping it to meet modern problems" (p. 22). Consequently, Buchanan has inspired his then Carnegie Mellon colleague David Kaufer to place rhetoric under the aegis of design (Kaufer & Butler, 1996). During the 1990s, this design-as-rhetoric trajectory continued to garner interests from rhetoricians who studied extra-textual arguments, including David Fleming (1996, 1998a, 1998b) and Kostelnick (1995, 1996), before being sidetracked by the dot-com phenomenon and the World Wide Web invention at the turn of the century.

For technical communication scholars, the effort to create an explicit connection between design and writing is a critical juncture—where *product* meets *process*—to forge a more holistic approach for communication (cf. Miller, 1985; Geisler, 1993; Medway, 1996a, 1996b; Winsor, 1996; Lewis, 1999). However, the concept of design as a solution for writing problems did not receive the uptake that rhetoric scholars in the 80s and 90s had hoped. It wasn't until 2009, when Richard Marback again offered design as a "new" paradigm for composition. Marback (2009) noted that the "centrifugal forces of critique in composition studies are giving way to centripetal interest in design, reinvigorating practical interest in agency" (p. 398). Thanks to the increased attention given to multi-modality and multimodal composition, writing studies as a whole has become more accepting of design approaches to composing, especially when it involves multimedia technology and situations that require solutions beyond text-only mediation. While Kostelnick (1989) sought to reinvigorate the process paradigm, Marback (2009) focused on the writer's agency and considered design—specifically design thinking—as a way to confront the *wickedness* of writing.

This wickedness is drawn from the notion of "wicked problems" coined by design theorists Horst Rittel and Melvin Webber (1973), which refer to societal problems that are complex and situationally unique in formation; lacking in linear explanations; difficult or impossible to solve with testable and finite solutions; and contagious in that any potential solution spawns other wicked problems. According to Rittel and Webber, wicked problems share the following characteristics:

1. They do not have a definitive formulation.
2. They do not have a "stopping rule." In other words, these problems lack an inherent logic that signals when they are solved.
3. Their solutions are not true or false, only good or bad.
4. There is no way to test the solution to a wicked problem.
5. They cannot be studied through trial and error. Their solutions are irreversible so, as Rittel and Webber put it, "every trial counts."
6. There is no end to the number of solutions or approaches to a wicked problem.
7. All wicked problems are essentially unique.
8. Wicked problems can always be described as the symptom of other problems.
9. The way a wicked problem is described determines its possible solutions.
10. Planners, that is those who present solutions to these problems, have no right to be wrong. Unlike mathematicians, "planners are liable for the consequences of the solutions they generate; the effects can matter a great deal to the people who are touched by those actions."

In "What Can Design Thinking Offer Writing Studies," James Purdy (2014) argued that "design thinking offers a useful approach for tackling 'wicked' multimodal/multimedia composing tasks" (p. 614). Purdy contended that design thinking forces writing studies to move beyond print based conditions and explore other modalities as available means of meaning making. "Invoking design," Purdy wrote, "can serve to answer Jody Shipka's call for the discipline to focus on all communicative practices, not just writing" (2014, p. 632). In the same year, Carrie Leverenz (2014) also advocated for design thinking as a teaching framework and composing process for multimodal texts: "it eliminates the question of how to fit multimodal composing into writing classes since it focuses on designing solutions to problems rather than creating forms for their own sake" (p. 3). It is safe to say that design thinking is achieving a steady state in composition scholarship given its connection to multimodality. As Kelli Cargile Cook (2002) argued, multimodal rhetorical skills encourage "students to understand and be able to analyze, evaluate, and employ various invention and writing strategies based upon their knowledge of audience, purpose, writing situation, research methods, genre, style, and delivery techniques and media" (p. 10). These skills are crucial for students to be successful, agile technical and professional communicators today. Moreover, as Jody Shipka (2005, 2011) demonstrated, design thinking and multimodal literacy can also help bridge the gap between academia and workplace through the varied communicative and composing practices students engage in the classroom that may also be performed in the workplace, such as websites and multimedia presentations. Scott Wible (2020) reinforced the argument that design thinking lets students develop creative habits of mind that can help address multidimensional problems.

Yet, it was not until more recently that design thinking began to appear in journals and conference proceedings specific to technical communication:

- Writing to the broader professional communication community, Ann Hill Duin and her collaborators from the University of Minnesota showcased design thinking as a methodology for team organizing (Duin, Moses, McGrath, Tham, & Ernst, 2017).
- To participants of the Association for Computing Machinery Special Interest Group on Design of Communication (SIGDOC) annual conference, Liz Lane (2018) showed that the design thinking process can be useful in scaffolding service learning projects.
- In *Programmatic Perspectives*, journal of the Council for Programs in Technical and Scientific Communication, Jennifer Bay, Richard Johnson-Sheehan, and Devon Cook (2018) posited design thinking as a strategic pedagogical approach for developing entrepreneurial competency in technical communication students.
- Following a well-attended panel at the 2018 Conference on College Composition and Communication, the *Journal of Business and Technical Communication* dedicated a special issue on design thinking approaches edited by Rebecca Pope-Ruark, myself, Joe Moses, and Trey Conner (Pope-Ruark, Tham, Moses, & Conner, 2019).
- Through the lens of game studies, Laquana Cooke, Lisa Dusenberry, and Joy Robinson (2020) argued in *Technical Communication Quarterly* that design thinking acts as a viable framework for increasing students' ability to solve macro- and micro-level problems in technical communication.

At a time when technical communication is undergoing an identity shift from traditional documentation and communicating complex information to content strategy and user experience (UX) design (cf. Redish & Barnum, 2011; Verhulsdonck, Howard, & Tham, forthcoming), design thinking serves as a critical component in mediating this process of change. It provides a language for understanding the work of technical communication as problem solving. Along the same vein, it helps technical communicators to identify as designers— designers of information, designers of user experience, designers of technological solutions.

However, because resources are scarce and popular sources like blogs and business forums are inconsistent with their use of the term, design thinking is often mistaken for a design exercise focusing on visual or graphic design. Some have treated design thinking as a one-size-fits-all formula with a linear, step-by-step process to devise solutions. This book seeks to rectify these misconceptions, starting with a historical account of design thinking.

Where Did Design Thinking Come From? A Brief Historical Sketch

Since the mid-2000s or so, the term *design thinking* has gained growing traction in the business world as well as the creative and computing fields. Thanks to mentions in trade publications like the *Harvard Business Review, Wired* and *Forbes*, design thinking has become not just a corporate buzzword but also piqued the interest of scholars who work at the intersection of technology development, software engineering, and social computing, among other industries. While some have argued that architectural studies should "own" design thinking ("The architecture of," 2019), since two of the most influential books on design thinking were authored by architecture theorists Bryan Lawson (1980)—*How Designers Think*—and Peter Rowe (1987)—*Design Thinking*. However, that's only part of the history of design thinking development. Design thinking is in fact a result of evolution from concepts and methodologies originated in fields of industrial design and engineering, social and computer sciences, and product development.

Historians have traced the origin of design thinking to the advancement of product design methodologies in two different (almost opposing) directions pursued by designers in the 1950s–1960s. One of the directions was led by inventor Buckminster Fuller at MIT. Fuller coined Design Science as a systematic approach to design through effective application of science. His methods were process-driven, but most importantly they were performed by experts from

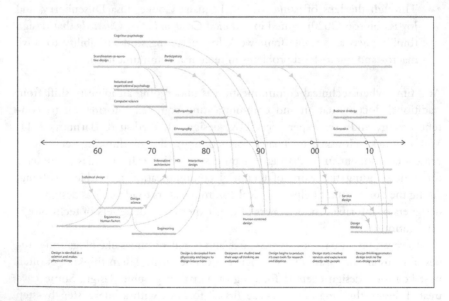

FIGURE 1.1   Key developments of design thinking

Source: Designed by Jo Szczepanska. Used with permission.

varying backgrounds and with specialized ways of working. This made Fuller's design teams multidisciplinary, a core characteristic of modern design thinking. Nevertheless, Fuller's approach was elitist and difficult to replicate in common businesses because he only enlisted individuals from the best universities and labs.

While Fuller was assembling his dream team in the US, designers in Europe took a much different route to product design. Scandinavian designers from well-known projects like Utopia and DEMOkratiske were experimenting with what they called cooperative design, or co-design methodology. Instead of hiring only experts to design teams, this participatory design approach aims to be non-selective and invites anyone who are interested in co-designing products and services to participate in the design process. This methodology has a great influence on the service design movement today that relies on creative, action-based design practices like mock-up envisionment and cooperative prototyping, among others. Arguably, the Scandinavian approach is also the root for human-centered design today. One distinction between this approach and the expert-based design approaches in the 1960s western world is user empowerment—the Scandinavian methodology sought to include the voice of actual users in the design process instead of making assumptions about user needs.

At the same time when aspiring designers were studying both the popular American and Scandinavian design models, Austrian-American designer Victor Papanek emerged as a strong advocate of socially and ecologically responsible design in the 1970s. Papanek was known for his integration of anthropological philosophy in his design practices and has contributed to the cross-disciplinary movement in product design.

As mentioned earlier, Rittel and Webber (1973) coined the term "wicked problem" from their examination of phenomenology as a philosophical approach to design studies. It wasn't until Buchanan's (1992) "Wicked Problem in Design Thinking," that the term became popular for designers. Buchanan (re)invigorated the discussion of wicked problems as a driving force for innovative design. Notably, the 90s was an important decade for design thinking not just because of Buchannan but also due to a fashionable three-way merger. In 1991, David Kelly (professor at Stanford), London-based designer Bill Moggridge, and Mike Nuttall of Matrix Product Design merged their companies to form IDEO, a multinational design and consulting firm. Between the 90s and early 2000s, IDEO had attracted influential talents from both academia and industry.

Alongside the rapid evolution of personal computers and mobile devices, as well as radical designers at Apple, Xerox, and IBM, IDEO managed to popularize design thinking and launched educational programs via the Stanford Hasso Plattner Institute of Design, fondly known as the Stanford d.school. Long-term members of IDEO, David Kelly and Tom Kelly also wrote best-selling books on design thinking and creative confidence. Tim Brown, an industrial engineer turned IDEO's CEO, has become the strongest advocate for design thinking and brought its ideals into the 21st century.

There are many figures worth noting in the design thinking timeline, but perhaps the most critical is a designer who has returned to the root of design and its purpose. With Nigel Cross, Kees Dorst argued that wicked problems cannot rely on rational problem-solving methods (Dorst & Cross, 2001). Dorst (2006, 2011, 2015) went on to suggest that designers should use a rhetorical (discourse) framing method to guide their innovation. Next to Buchanan, Dorst is likely the most relevant theorist of design thinking to rhetoric and technical communication scholars. As Scott Weedon (2019) summarized, "Dorst's framing approach provides technical communicators a method to analyze the various discourses and practices that constitute a situation in order to invent an innovative frame" and "technical communicators can use their rhetorical knowledge to invent productive metaphors that suggest tactical implications for adequately addressing complex situations" (p. 429).

Into the second new decade of the 21st century, design thinkers are also more concerned about social issues than ever. A key figure in social design, Deborah Szebeko, founded the British design agency ThinkPublic, and has been an advocate for design innovation within the public and non-governmental organization sector. Szebeko's work has a strong influence on the social justice movement many design thinking practitioners have confided in today.

Even with just a brief overview of the history, we can see that design thinking can be traced to multiple disciplines and it is an amalgamation of design practices. Design thinking has been contextualized by the shifting practices led by various design industries. The key figures highlighted in this short historical account have each enlightened designers on how to *think about design* and the different ways to *do design*. In the next section, I describe the common methodology for design thinking as an embodiment of these conceptualizations.

The Design Thinking Mindset and Methodology

Focusing neither only on design nor thinking, design thinking is a combination of a methodology and mindset for innovative problem solving. It forwards a problem-based approach to innovating solutions by offering guiding principles for choosing and using various methods to understand problems and users. However, as human–computer interaction researcher Nigel Cross (1982) put it, design thinking practitioners should first and foremost think like a designer. Although design thinking seems like a particular framework for creating designed solutions, Cross warned designers about the perils of "scientizing" design (Cross, 2001). Cross believed that design can be organized and somewhat systematic, but it should not profess any positivist doctrine. Designers can use the many scientific underpinnings of design—like material science, engineering science, and behavioral science—to understand design problems and perform design, *design thinking is not a scientific method*. In other words, design thinking does not prescribe any methodological rules for innovation.

Stefanie Di Russo (2016), former senior consultant of design strategy at Deloitte Australia, defined design thinking as

> a term widely used outside of the design industry to describe the innovative and human-centered approach used by designers in their practice. . . . [It] has erupted outside of design practice as a new approach for innovation and transformation, piquing the interest of leaders from business, education, government, through to not-for-profit organisations.
>
> *(p. 3)*

In an earlier blog post, Di Russo (2012) chronicled the development of design thinking, noting especially the significant contribution to design thinking made by UX guru Don Norman. In *The Design of Everyday Things* (revised edition), Norman (2013) stated:

> Designers resist the temptation to jump immediately to a solution for the stated problem. Instead, they first spend time determining what basic, fundamental (root) issue needs to be addressed. They don't try to search for a solution until they have determined the real problem, and even then, instead of solving that problem, they stop to consider a wide range of potential solutions. Only then will they finally converge upon their proposal. This process is called design thinking.
>
> *(p. 219)*

Per Norman's view, design thinking facilitates a problem-based mindset—a designerly way of thinking that Cross advocated. For some designers including Di Russo, design thinking should cultivate transformation. In *Design-Driven Innovation*, Roberto Verganti (2009) articulated the strategy of design thinking as one involving radical change. In particular, the process of design-driven innovation involves listening to interpreters or what Verganti referred to as "forward-looking researchers who are developing, often for their own purposes, unique visions about how meanings could evolve in the life context we want to investigate" (p. 13). Verganti's approach shifts the emphasis in design thinking from process to philosophy. Instead of merely providing a procedure for invention, Verganti's view of design thinking concentrates on an ideal that pays attention to the design mindset.

But how exactly does design thinking *work*? As a methodology informed by the aforementioned mindset, design thinking typically manifests in a solutions-generating process involving five phases: empathize, define, ideate, prototype, and test, represented by Figure 1.2.

Empathy is the foundation of design thinking. It serves as the first guiding principle for any design thinking-powered process. To empathize is to understand users and stakeholders' behaviors in the context of their lives, and to engage with

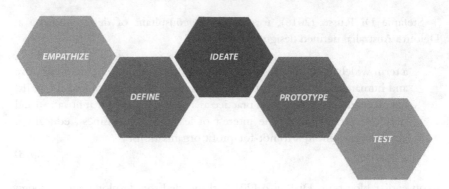

FIGURE 1.2 The basic model of design thinking adapted from the Stanford model (Stanford d.school, n.d.)

them directly in order to discover their needs, motivations, efforts, and other stories pertaining to the lives of those affected by a particular problem or situation. There are many ways to foster empathy in designers, including contextual inquiry, photovoice, diary study, bodystorming, and journey mapping.

As Norman put it, effective design gets to the root of a problem. **Definition** is a mode in design thinking for unpacking and synthesizing initial exploration findings into compelling insights and specific scope of challenge. This step in the design process helps designers to develop a deep understanding of user requirements and craft focus, actionable problem statements that guide the design direction. A strong definition also inspires and empowers the design team to pursue innovative ideas in response to wicked problems. Popular methods for defining problems include synthesizing findings from empathy maps, asking "How might we?," and generating the designer's point of view.

Ideation is a set of exercises that generate radical design alternatives. Unlike the definition mode, ideation is a mode of flaring rather than focus. It seeks to cast a wide net to catch a large quantity of wild ideas before narrowing them down. Designers should aim to go beyond obvious answers and use the previously crafted problem statements to create unexpected solutions. Diversity of ideas is celebrated at this stage. Teams should harness their collective imagination power and actively include varying perspectives to the problems at hand. Exercises that can help to spark innovative ideas include storyboarding, dot voting, affinity mapping, and the four-category method.

Prototyping is the "make it real" phase of the design process. It is about getting ideas out of the designers' head into the real world. While the Stanford d.school's guidelines state that prototypes should take a physical form, I argue that digital manifestations of design ideas are just as valuable as 3D object-based prototypes. Nonetheless, design teams should leverage the affordances of rapid-prototyping tools like computer-numerically controlled (CNC) milling, digital interface builders, and low-fidelity methods like paper-based wireframes or even post-it notes to create quick displays of the selected design solution.

With the prototyped idea, teams can further explore the viability, desirability, and usability of their design through the **testing** phase. The purpose of testing is to gather feedback on the designed solution and engage those who might be affected by the proposed solution. The methods associated with this phase allow designers to test their point of view, learn about user behaviors, and identify ways to refine the solution using user feedback. Standard usability testing methods are appropriate for this phase.

In the Design Thinking Methods and Exercises appendix (Appendix B) of this book, I provide further descriptions of exercises and methods mentioned in these five phases. Examples of applying some of these methods can be found within my case studies in the following chapters.

User- and Human-Centered Design

Design teams can determine when and how to journey through the five basic phases of the design thinking process. Regardless of a team's preferred workflow, an important consideration is that design thinking should not be treated as a rigid, linear design process. Teams should not treat the five phases merely as checkpoints to reach. Instead, a design thinking process should exercise iterative design. Different factors including new ideas, user feedback, and technological opportunities or limitations can take a team from one phase back to a previous phase or more. This should be embraced as an integral process to design so as to ensure user-centered innovations.

As design thinking prioritizes the user, technical communication values user-centered design as one of its principal practices. User-centered design was born of usability studies. Don Norman and Stephen Draper (1986) coined the term in their book, *User-Centered System Design: New Perspectives on Human-Computer Interaction*, introducing it to professions like human factors, software engineering, and technical communication. The rhetorical roots of technical communication and its existing investment in audience analysis helped the field adopt user-centered design in the 1990s. Notably, Robert Johnson (1998) provided a theoretical model that combined rhetoric and user-centered theory. Johnson's model rationalizes the adaptation of user-centered design in technical communication pedagogy. It resulted in the addition of courses like usability and user experience courses in technical communication curricula (Breuch, Zachry, & Spinnuzi, 2001). Another sign of this adaptation can be seen through the identity evolution in technical communication programs. For example, the Department of Technical Communication at the University of Washington—Seattle became the Human Centered Design and Engineering department in 2009 to reflect the prominence of user-centered design in technical communication (Department history, n.d.). Elsewhere, like the Technical Communication and Interactive

Design departments at Kennesaw State University and Metropolitan State University (Minnesota), and the Technical Communication and Information Design program at University of Colorado at Colorado Springs have added "design" to their core identity as a way to emphasize its place in technical communication instruction.

There have been discussions and debates regarding the use of the word "user" vs. "human" as in human-centered design instead of user-centered design. Citing Rob Kling and Susan Leigh Star (1998) and William Rouse (2007), Mark Zachry and Jan Spyridakis (2016) noted that human-centered design focuses on the social dimension of user interaction with systems that user-centered design has sometimes overlooked. Simon Baron-Cohen (2011) criticized that by making the human element invisible, user-centered design dehumanizes the user in interactive system experiences. Marina Yalanska (2018) of FAQ Design Platform (tubikstudio.com) took on the slight nuances in the two terms and observes that "human-centered design is the process of things deeply based on general natural characteristics and peculiarities of human psychology and perception" while "user-centered design is [a] more focused and concise version of human-centered design with deeper analysis of target audience" (n.p.). Yalanska summarized that human-centered design and user-centered design work hand-in-hand; the idea is to first design for humans, then define the needs of the specific category of users. Nevertheless, the shared missions of user-centered design and human-centered design have left the semantic bale unresolved. In this book, I apply the two terms interchangeably, with specific use of human-centered design when discussing the humane aspects of innovations.

Design thinking supports user- and human-centered design. It shares the same belief that the user—and empathy and understanding for the user—should be the primary motivation in all design decisions. The iterative nature of design thinking considers the *whole* user experience; it takes into account users' hopes and needs (through empathy and definition), as well as their actual reactions and behaviors (especially during the testing with prototypes).

But Why Making?

Now, one may be surprised to see the inclusion of "making" in this book as the subject matter is design thinking in technical communication. "Making," as popularized by a recent development called the Maker Movement (which I discuss further in the next chapter), has entered the field of technical communication, in practice and pedagogy. Within computers and writing—a subfield of writing studies that has tremendous influences on technical communication—scholars have given attention to what making can teach them about learning and multimodal delivery. Making was a keynote subject for the annual Computers and Writing conference given by James Paul Gee (2013), and continues to manifest in conference programs, journal articles, and books. Chet Breaux (2017) observed that many writing scholars are already very interested in the practices used by makers and the artifacts they create even though the terms "maker" and "making" were popularized only recently. Amid differing threads of discussions and converging interests, I see an opportunity to create a space for patching "making" and technical communication together through design thinking, and weaving existing theories of multimodality and multiliteracies to create a case for meaningful making in technical communication pedagogy.

Besides the rationale proposed by Breaux (2017), this move is also largely inspired by Joyce Locke Carter's Chair's Address, "Making, Disrupting, Innovating," delivered at the 2016 Conference on College Composition and Communication (CCCC) in Houston, Texas. Using real-time voice sensing, motion tracking, and corresponding visual displays, Carter performed a sensational address that demonstrated a potential future for multimodal rhetoric that involves not just textual or auditorial appeals but also imagerial and gestural. Besides its demonstrative effects, Carter's address called writing studies scholars into valuing making as a valid and plausible way of learning in the 21st century. Carter's exigence is built upon the historical impact that making has on our field and its advancement. She called our attention to several innovative instances, such as Daedalus, ELI Review, BABEL, and EyeGuide, all of which have helped define writing studies as a productive discipline that contributes to the betterment of our knowledge society.

> When I talk about making, I'm flipping the power and flipping the epistemology, and saying that when you make, you dictate what will happen. You create new things that hopefully challenge the status quo (which is also the goal of advocacy), and while some, if not most, efforts end in failure, some will be quite disruptive. (Carter, 2016, p. 390)
>
> And when you hear one of your colleagues say the words "my startup company," or "my new app," you might be tempted to think, "Oh, that's a bit unusual for someone at the C's to talk like this." I'll argue that that kind of statement at the bar, or in a session, should be seen as desirable and normal—as normal as someone who mentions "my new book," or "my research" or "my advocacy."
>
> *(Carter, 2016, p. 391)*

While Carter's address begged for more litigable theoretical frameworks for making, disrupting, and innovating in writing studies, her contention for a multimodal future of writing studies is well taken and can be used as a springboard to research that investigates design-driven and problem-based technical communication pedagogy. Along with Carter's motivation, making responds to our field's growing needs for viable pedagogical frameworks to integrate multimodal composition and design thinking with technical communication programs. As I iterate throughout this book, making subscribes to a particular characterization of technical communication set out by Johndan Johnson-Eilola and Stuart Selber (2013):

> Technical communicators do not merely learn skills; they must also learn how to learn new skills, upgrading and augmenting their abilities as they mature in careers, analyzing the matches and mismatches between what they currently know and what a communication situation demands. . . . [They] must learn to become reflective problem solvers.
>
> *(p. 3)*

Johnson-Eilola and Selber considered problem-solving to be a productive characterization for it acknowledges the extent to which our field contributes to technological development and its use, the interpretation of rhetorical situations, and the design of viable solutions based on context, complexity of the tasks and their characteristics. With Johnson-Eilola and Selber's characterization, I address the core question Leverenz (2014) asked about multimodal writing instruction, "How can we teach writing so that students learn to use words and other language resources to define and respond in creative ways to problems they see as important?" (p. 4). I take this challenge a notch higher by asking how we might deliver technical communication instruction such that writing becomes the "head fake[2]" (Pausch & Zaslow, 2008) in the learning process, so that other desirable traits—like greater rhetorical awareness, collaboration skills, critical thinking, ethical decision making, etc.—might emerge as the learning outcomes.

Across many fields, particularly education, engineering, and business, the Maker Movement and its philosophy have been adopted for creating innovative, open, and collaborative communities of learners and makers (Hagel, Brown, & Kulasooriya, 2013; American Society for Engineering Education, 2016). In these communities—fondly known as makerspaces—students cultivate a strong sense of agency in solving problems they identified as important through a design thinking process. This process requires students to combine creative and analytical approaches to define problems and invent desirable solutions by collaborating with others. Although not limited only to schools and university campuses, most makerspaces today are built within an education setting (Carlson, 2015). The idea of a Maker Education is to create such spaces where students collaborate across disciplines and tackle complex problems.

Many technical communication scholars come from a background in writing studies and may be curious where making and professional/technical writing overlaps. While it may seem far-fetched at first, there are a number of recognized connections between making and writing. For one, making, much like writing, is process-oriented; it involves the drafting and composing of an artifact, trial and error, revision, and reflection (Sheridan, 2010; Gierdowski & Reis, 2015). One might also recognize the similarities between making and the growing conversation about multimodality that is taking place in the areas of rhetoric and composition, computers and writing, and digital rhetoric. In writing studies, we have been challenged to reconsider what we see as "texts," as James Porter (2002) pointed out, "We are already in the age of new media, where visual and video forms of expression supersede alphabetic text" (p. 389). The material outcome of a "making" could very well communicate a message the same way as conventional alphabetic texts, if not more effectively. What's more, the project-focused, process-oriented maker approach that favors co-creation is comparable to the participatory, collaborative knowledge-making practices that are highly regarded in rhetoric and composition studies (i.e., Ede & Lunsford, 1990; Yancey & Spooner, 1998). Finally, emphasizing the importance of translating knowledge into practical application, Miles Kimball (2017) stressed the need to leverage the technical communication classroom for change-making:

> Compositionists are rethinking general college writing instruction, for example by having students write "multimodal" compositions. . . . This change reflects a growing awareness of the importance of technology in human communication. Multimodality, however, does not always emphasize instrumentality; many multimodal compositions are simply expressive writing in multiple media. We owe all students an opportunity to learn how to communicate in a technological world—not just by writing a multimodal essay instead of a lexical essay, but by learning to use technologies of communication to bring about practical change. Technical communication is ideally situated to help do just that.
>
> *(p. 350)*

Given these observations, I see an opportunity to identify the viability of a maker pedagogical approach for technical communication, through the perspectives of design thinking.

Addressing Wicked Problems in Technical Communication Pedagogy

As I have forecasted in the opening of this chapter, technical communication pedagogy faces challenges in staying relevant to industry trends and market needs, resulting in a need for the field to reinvigorate its pedagogical frameworks to include current methodologies and philosophies such as design thinking in technical communication programs. Here, I trace the cause to these challenges by looking at our resistance to new technology (mainly out of fear and discomfort), the rise of multimodal composing tools, and the "wicked" difficulty in delivering multimodality and design-centric learning in technical communication courses.

Any envisionment of a future technical communication instruction needs to recognize a major shift in how we compose and consume texts in this age of information technology. Through advancing web platforms, social media, analog and augmented realities, and other virtual interactive tools, writers are moving beyond using just alphabetic texts to access information and communicate with others. Yet, in our classrooms, many instructors still resist teaching with new technology for various reasons (Hickey, 2000; Hart-Davidson, Cushman, Grabill, DeVoss, & Porter, 2005; Kemp, 2005; Knievel, 2006; Hewett, 2015). In technical communication, we still question whether we should teach specific technology (Garrison, 2018). Bonita Selting (2002) addressed this question by surveying ATTW members regarding their roles as teachers of technical writing in relation to demands to also teach technology skills, concluding that "technological determinism—shown by a tendency to turn a technical communication course into a software tools course—can be seen as a threat to effective teaching of complex workplace rhetoric" (p. 251). In addition, our discipline often aligns with a view reticent toward teaching tools: Reporting on behalf of the College Composition and Communication Conference Committee for Effective Practices in OWI (Online Writing Instruction), Beth Hewett shared OWI Principle 2: "An online writing course should focus on writing and not on technology orientation or teaching students how to use learning and other technologies" (2015, p. 45).

Such resistance has led to challenges in infusing up-to-date tools and digital literacies into our pedagogy, including understanding and producing multimodal texts. Aaron Doering, Richard Beach, and David O'Brien (2007) argued that given the ready access to Web 2.0 tools and worldwide audiences, we need to infuse multimodal and digital literacies into writing instruction so students could learn to use media tools to "effectively attract, engage, and influence their audiences," and "foster constructivist, inquiry based learning related to fostering critical thinking" as well as effective writing/communication practices (pp. 41–42). Karl Stolley in his "Lo-Fi Manifesto" (2008, 2016) encourages writing instructors to assume such responsibility:

> Those who teach have an even more pressing responsibility to learn and then engage students with digital approaches and technologies that students themselves would not likely discover independently. Students must be afforded the opportunity to write markup, programs, APIs, and commit messages in the same range of learning situations as they write essays and exams today. They must be encouraged, supported, and even joined by their instructors in failed first efforts. The richest learning experiences reveal how failure and crude initial work transform to something better only through ongoing research and revision.
>
> *(2016, n.p.)*

Earlier, Marback (2009) argued that the concept of design can be appealing to writing studies, particularly for those "teaching writing in digital media" (p. 397), and Leverenz (2014) consider multimodal/multimedia composing as "wicked" tasks that require design thinking as a generative or productive approach to the composing process. Purdy (2014) observed that writing studies programs are, institutionally speaking, moving closer to be associated with design disciplines due to the growing demand to teach information design, writing for new media, and visual rhetoric or communication. Scholars from computers and composition as well as technical communication would agree that traditional writing instruction does not always fit the needs of these new domains.

Like Leverenz and Marback, I consider technical communication pedagogy a "wicked problem" beyond its procedural complication (Marback, 2009, p. 400) into how instructors could teach students to move across and beyond linguistic resources to solve communicative problems they identify and consider as important, in innovative and effective ways. Such wickedness requires us to treat the technical communication classroom not just as a site for information delivery and proficiency testing, or merely a place to practice producing various genres, but a space for practical guidance—through instructor facilitation and peer support—to solve technical communication problems through direct experience with tangible materials. This is particularly important for teaching in an information age, where students are equipped with cutting-edge tools and inventive methods that allow them to create content with ease and efficiency. For instructors, this poses new challenges in terms of fostering rhetorical awareness as well as technical knowledge in students such that they are able to utilize all available means of communication to achieve their goals.

What Can Design Thinking and Making Offer Technical Communication Pedagogy?

Johnson-Eilola and Selber (2013) considered problem-solving as a productive characterization for it acknowledges the extent to which our field contributes to technological development and its use, the interpretation of rhetorical situations, and the design of viable solutions based on context, complexity of the tasks and their characteristics. Making and design thinking offer a social and materialist dimension to such problem solving in technical communication. To date, most literature cites constructionism and constructivism as underlying principles for maker and design based instruction (Donaldson, 2014; Vaughn, 2017). Constructivism is more prominent in writing studies history, which is grounded in the works of Jean Piaget (1952, 1957, 1973), Lev Vygotsky (1978a, 1978b), and Jerome Bruner (1960, 1966, 1996). A constructivist approach to epistemology holds that meanings are created based on our constant interactions with the physical, mental, and social worlds we inhabit, and we negotiate those meanings by building and adjusting our internal knowledge structure and organizing our perception and reflection on reality (Swan, 2005). Whereas constructivism is a theory of knowledge that sees learning as an active, social process in which students reconstruct knowledge rather than simply receive a transmission of knowledge from a teacher, constructionism is a theory of learning that suggests that the internal construction of knowledge is most readily achieved when the student is also engaged in the active construction of a personally meaningful and tangible product (Papert, 1980, 1993). For constructionists, emphasis is put on creating and discovering, and tapping into the learner's natural inclinations toward problem-solving.

Design thinking and making are pedagogical efforts that involve creating opportunities that let students attempt to solve problems that are complex in nature. The constructivist-constructionist nature of making and design thinking acknowledges learning as something that comes from the learner's question or impulse and is not imposed by the instructor. This empowers learners to connect with everything they know, feel, and wonder, stretching themselves into learning new things, and liberating them from their dependency on being taught (Blikstein, Martinez, & Pang, 2015). Both design thinking and making also stress that students learn best by creating tangible objects through authentic, real life learning opportunities that allow for a guided, collaborative process which incorporates peer feedback. A signature activity in such process is fondly known as the "design challenge" (described in the next chapter), which typically revolves around complex social problems that requires the participants to work in cross-functional teams and exercise the design thinking methodology—empathizing, defining, ideating, prototyping, and testing. We have affirmation from industry practitioners that design challenges can promote collaboration and critical problem-solving, and are becoming a staple creative activity in the UX profession ("Why you should," 2018).

The material dimension of situated problem solving has been a notable focal point in writing studies scholarship (Haas, 1996; Haas & Witte, 2001; Rifenburg, 2014). Programmatically speaking, design thinking and making enable a "multimodal pedagogy" that Andrew Bourelle, Tiffany Bourelle, and Natasha Jones (2015, p. 309) forwarded. Taking heed from Claire Lauer (2009), who contended there must be "an understanding of design-as-process and the situated choices and strategies students need practice developing" (p. 237), Bourelle et al. (2015) showed that a multimodal approach to technical communication pedagogy can prepare students for future industry projects. By emphasizing materiality and rhetorical choices in the problem-solving process, making and design thinking in the technical communication classroom may achieve these goals.

A Note on Methods

The three succeeding chapters feature multiple case studies that used mixed methods—e.g., interviewing, survey, ethnography, and collaborative autoethnography—for data collection and analysis. Whenever available and appropriate, I include verbatim quotes from participants of my studies to draw attention to their voices. As a qualitative researcher, I am most invested in identifying emergent points of interest through particular instances rather than making generalizations through large data sets. The findings from the ensuing studies provide unique insights about the affordances of design-centric practices in pedagogy and the profession.

Each highlights an aspect of design and making that could enrich our existing work. Together, they help paint the bigger picture of design thinking approaches in technical communication. Readers may notice that I do not draw specific conclusions from the studies—whether we should or should not employ certain practices—because that is not the goal of this project. Instead, I am motivated to reveal what happens when we integrate design thinking with our classroom and workplace, and to offer suggestions for those who are interested in experimenting with such integration. Afterall, design thinking celebrates a pioneering spirit and embraces both success and failure.

Summary and Takeaways

This chapter showed the trajectory of design-centric and design thinking influences in our field's scholarship. The chapter provided a quick overview of design thinking's history and described the key phases that make up the problem-solving methodology and mindset. In addition, it demonstrated the relationship between design thinking and user-centered design. The introduction to "making" as an active design and problem-solving strategy showed its potential to address our current technical communication pedagogical needs and its overlaps with the design thinking framework. Key takeaways from this introduction chapter are:

- Increased attention to multimodality in writing studies contributed to the turn to design as an integrated (process + product) approach to teaching and learning.
- Design thinking is an amalgamation of practices informed by multiple disciplines, with a noticeable origin from design science and the Scandinavian co-design models in the 1950s and 1960s.
- Design thinking supports iterative and user-centered design, core principles in technical communication.
- Making is a recent popularized educational activity. Powered by design thinking, making can address the material and multimodal issues facing technical communication pedagogy.

Learning Activity: A Design Thinking Orientation

A rather quick exercise (about 50 minutes) to understand the design thinking mindset and methodology is through a rapid orientation that I love conducting at the start of a new class. The preparation for this exercise is minimal: The facilitator (instructor, trainer) just needs to provide some readily available crafting materials like the following, or anything they might find from a recycling bin:

- Adhesives: glue sticks, tapes, stickers
- Cutting tools: scissors, pen knives

- Canvas: scrap papers, color papers, card boards, paper boxes
- Drawing utensils: color pens, crayons, chalks
- Random items: ribbons, sticky note pads, popsicle sticks, tissue boxes

First, put your participants (in my case, students) into pairs, preferably with someone they haven't known too well. Then, have them spend just about 1 minute introducing themselves to each other. The next step is to announce the goal of the orientation. One that I like to use is called "Reinventing the Classroom Experience" because it is usually comprehensible by my students.

In their pairs, Student A will spend 3 minutes interviewing Student B about their everyday classroom experience. This should be an open inquiry without specific focus on an aspect of the classroom. The interviewer should focus on listening and not notetaking. Once the time is up, the interviewer has 45 seconds to jot down the main takeaway from the interview. Then, vice versa, Student B interviews Student A, and writes down their takeaways. This segment of the activity represents the **empathy** mode of design thinking.

The next step is to **define** the problem. Both students spend 2 minutes to write a point-of-view (POV) statement using the following structure:

_____ (name of interviewee) _____ needs
_____ (required service, things, spaces, etc. that's currently missing) _____
in order to _____ (achieve desired goals) _____.

An example would be:

Student A needs
a *movable, wireless charging port*
in order to *keep her laptop charged during class.*

At this time, the partners will exchange the POV statements and ask for feedback. Changes can be made during this exchange to create a more accurate statement. I also ask students to observe their classroom setting (and beyond, thinking about the campus in general), and to pay attention to ways we interact with objects and spaces.

Now, enter the **ideation** phase. Using the POV statement and observation of classroom spaces, students were asked then to generate at least 5 radical solutions to meet their partner's needs. They are given only 8 minutes to do so. To throw in an incentive, I give a small prize (bag of candies) to an individual who came up with the most ideas within the time frame.

Once the time is up, students would take turns to share their radical ideas with their partner, and spend a few minutes selecting one "big idea" to refine on paper and use it for the next part of the activity—**prototype**. Using the materials I supplied, students then spend the next 8–10 minutes building a model for their "big idea." They are encouraged to build their model as close to their ideas as possible.

When done, students presented their prototyped solutions to their partner. They were asked to focus on what worked, what needed to be improved, and to take questions from their partner. This is meant to simulate the **testing** mode. If time permits, I would let students iterate their prototype based on the feedback they received from their partner.

That is the end of this orientation activity. Students may keep their prototype if desired. In the next class period, I usually ask them to reflect on the following questions as a segue into discussions of user-centered design:

- How is design thinking similar or different from "traditional" research processes?
- How did you approach your partner's problem(s)?
- How did you come up with your "radical" solutions?
- What does your partner think about your solution?
- How did you modify your solution based on your partner's feedback?

Notes

1. Although "design" has appeared in composition studies literature as early as the 1940s, Melanie Yergeau (2015) noted in *Keywords in Writing Studies* that these explorations such as Arthur Minton's (1941) "Design for Composition" focused mainly on course planning and research methodology.
2. Randy Pausch, in his infamous talk, "The Last Lecture" (and book with the same title), shared the notion of "head fake"—or indirect learning. It refers to a situation where someone believes they are learning about one thing (that they usually are not interested in), but are really learning about something different and beneficial to them. A head-fake example that Pausch shared in his talk was youth sports: Parents don't usually care much about their children learning the intricacies of the sports they play, but instead they want them to learn about desirable values like teamwork, perseverance, and sportsmanship.

References

American Society for Engineering Education. (2016). *Envisioning the future of the maker movement: Summit report*. Washington, DC: American Society for Engineering Education.

The architecture of design thinking. (2019). Practice of architecture. Retrieved from https://practiceofarchitecture.com/2019/03/29/the-architecture-of-design-thinking/

Baron-Cohen, S. (2011). *Zero degrees of empathy*. New York, NY: Penguin Group.

Bay, J., Johnson-Sheehan, R., & Cook, D. (2018). Design thinking via experiential learning: Thinking like an entrepreneur in technical communication courses. *Programmatic Perspectives, 10*(1), 172–200.

Bekins, L., & Williams, S. (2006). Positioning technical communication for the creative economy. *Technical Communication, 53*(3), 287–295.

Blikstein, P., Martinez, S. L., & Pang, H. A. (2015). *Meaningful making: Projects and inspirations for fablabs and makerspaces*. Torrance, CA: Constructing Modern Knowledge Press.

Bourelle, B., Bourelle, T., & Natasha, J. (2015). Multimodality in technical communication classroom: Viewing classical rhetoric through a 21st century lens. *Technical Communication Quarterly, 24*, 306–327.

Breaux, C. (2017). Why making? *Computers and Composition, 44*, 27–35.

Breuch, L-A. M. K., Zachry, M., & Spinuzzi, C. (2001). Usability instruction in technical communication programs: New directions in curriculum development. *Journal of Business and Technical Communication, 15*(2), 223–240.

Bruner, J. S. (1960). *The process of education.* Cambridge, MA: Harvard University Press.

Bruner, J. S. (1966). *Toward a theory of instruction.* Cambridge, MA: Harvard University Press.

Bruner, J. S. (1996). *The culture of education.* Cambridge, MA: Harvard University Press.

Buchanan, R. (1985). Declaration by design: Rhetoric, argument, and demonstration in design practice. *Design Issues, 2*(1), 4–22.

Buchanan, R. (1992). Wicked problems in design thinking. *Design Issues, 8*(2), 5–21.

Cargile Cook, K. (2002). Layered literacies: A theoretical frame for technical communication pedagogy. *Technical Communication Quarterly, 11*(1), 5–29.

Carlson, S. (2015). The "Maker Movement" goes to college. *The Chronicle of Higher Education.* Retrieved from www.chronicle.com/article/The-Maker-Movement-Goes/229473

Carter, J. L. (2016). Making, disrupting, innovating. *College Composition and Communication, 68*(2), 378–408.

Cooke, L., Dusenberry, L., & Robinson, J. (2020). Gaming design thinking: Wicked problems, sufficient solutions, and the possibility space of games. *Technical Communication Quarterly.* Online first edition. Retrieved from www.tandfonline.com/doi/full/10.1080/10572252.2020.1738555

Cross, N. (1982). Designerly ways of knowing. *Design Studies, 3*(4), 221–227.

Cross, N. (2001). Designerly ways of knowing: Design discipline versus design science. *Design Issues, 17*(3), 49–55. Retrieved from http://oro.open.ac.uk/3281/1/Designerly-_DisciplinevScience.pdf

Department history. (n.d.). Human centered design & engineering. University of Washington. Retrieved from www.hcde.washington.edu/history

Di Russo, S. (2012). A brief history of design thinking: The theory. Retrieved from https://ithinkidesign.wordpress.com/2012/06/

Di Russo, S. (2016). Understanding the behaviour of design thinking in complex environments (Unpublished doctoral dissertation). Swinburne University of Technology, Melbourne, Australia.

Doering, A., Beach, R., & O'Brien, D. (2007). Infusing multimodal tools and digital literacies into an English education program. *English Education, 40*(1), 41–60.

Donaldson, J. (2014). The maker movement and the rebirth of constructionism. *Hybrid Pedagogy.* Retrieved from http://hybridpedagogy.org/constructionism-reborn/

Dorst, K. (2006). Design problems and design paradoxes. *Design Issues, 22*, 4–17.

Dorst, K. (2011). The core of "design thinking" and its application. *Design Studies, 32*, 521–532.

Dorst, K. (2015). *Frame innovation: Create new thinking by design.* Cambridge, MA: Massachusetts Institute of Technology Press.

Dorst, K., & Cross, N. (2001). Creativity in the design process: Co-evolution of problem-solution. *Design Studies, 22*, 425–437.

Duin, A. H., Moses, J., McGrath, M., Tham, J., & Ernst, N. (2017). Design thinking methodology: A case study of "radical collaboration" in the wearables research collaboratory.

Connexions: International Professional Communication Journal, 5(1), 45–74. Retrieved from https://connexionsjournal.org/wp-content/uploads/2019/12/duin-etal.pdf

Ede, L., & Lunsford, A. (1990). *Singular texts/plural authors: Perspectives on collaborative writing*. Carbondale, IL: Southern Illinois University Press.

Fleming, D. (1996). Can pictures be arguments? *Argumentation and Advocacy, 33*(1), 11–22.

Fleming, D. (1998a). Design talk: Constructing the object in studio conversations. *Design Issues, 14*(2), 41–62.

Fleming, D. (1998b). The space of argumentation: Urban design, civic discourse, and the dream of the good city. *Argumentation, 12*(2), 147–166.

Garrison, K. (2018). Moving technical communication off the grid. *Technical Communication Quarterly, 27*(3), 201–216.

Gee, J. P. (2013, June). Writing in the age of the maker movement. Keynote presented at Computers and Writing conference, Frostburg, MD.

Geisler, C. (1993). The relationship between language and design in mechanical engineering: Some preliminary observations. *Technical Communication, 40*(1), 173–176.

Gierdowski, D., & Reis, D. (2015). The MobileMaker: An experiment with a mobile makerspace. *Library Hi Tech, 33*(4), 480–496.

Haas, C. (1996). *Writing technology: Studies on the materiality of literacy*. Mahwah, NJ: Lawrence Erlbaum Associates.

Haas, C., & Witte, S. (2001). Writing as embodied practice: The case of engineering standards. *Journal of Business and Technical Communication, 15*, 413–457.

Hagel, J., Brown, J. S., & Kulasooriya, D. (2013). A movement in the making. Deloitte Insights. Retrieved from https://www2.deloitte.com/us/en/insights/topics/emerging-technologies/a-movement-in-the-making.html

Hailey, D., Cox, M., & Loader, E. (2010). Relationship between innovation and professional communication in the "creative" economy. *Journal of Technical Writing and Communication, 40*(2), 125–141.

Hart-Davidson, B., Cushman, E., Grabill, J., DeVoss, D. N., & Porter, J. (2005). Why teach digital writing? *Kairos, 10*(1). Retrieved from http://kairos.technorhetoric.net/10.1/coverweb/wide/index.html

Hewett, B. L. (2015). Grounding principles of OWI. In Beth L. Hewett & Kevin Eric DePew (Eds.), *Foundational practices of online writing instruction* (pp. 33–92). Fort Collins, CO: WAC Clearinghouse.

Hickey, D. (2000). Tangled up in blue: The web of resistance to technology and theory. *Academic Writing*. Retrieved from http://wac.colostate.edu/aw/papers/hickey

Johnson, R. R. (1998). *User-centered technology: A rhetorical theory for computers and other mundane artifacts*. New York, NY: State University of New York Press.

Johnson-Eilola, J. (1996). Relocating the value of work: Technical communication in a post-industrial world. *Technical Communication Quarterly, 5*(3), 245–270.

Johnson-Eilola, J., & Selber, S. (2013). *Solving problems in technical communication*. Chicago, IL: University of Chicago Press.

Kaufer, D. S., & Butler, B. S. (1996). *Rhetoric and the arts of design*. Mahwah, NJ: Lawrence Erlbaum Associates.

Kemp, F. (2005). The aesthetic anvil: The foundations of resistance to technology and innovation in English departments. In J. Carter (Ed.), *Market matters: Applied rhetoric studies and free market competition* (pp. 77–94). Cresskill, NJ: Hampton Press.

Kimball, M. (2017). The golden age of technical communication. *Journal of Technical Writing and Communication, 47*(3), 330–358.

Kling, R., & Star, S. L. (1998). Human centered systems in the perspective of organizational and social informatics. *Computers and Society, 28*(1), 22–29.

Knievel, M. (2006). Technology artifacts, instrumentalism, and the *Humanist Manifestos*: Toward an integrated humanistic profile for technical communication. *Journal of Business and Technical Communication, 20*(1), 65–86.

Kostelnick, C. (1989). Process paradigm in design and composition: Affinities and directions. *College Composition and Communication, 40*(3), 267–281.

Kostelnick, C. (1995). Cultural adaptation and information design: Two contrasting views. *IEEE Transactions on Professional Communication, 38*(4), 182–196.

Kostelnick, C. (1996). Supra-textual design: The visual rhetoric of whole documents. *Technical Communication Quarterly, 5*(1), 9–33.

Lane, L. (2018). Iteration for impact: Exploring design thinking & designing for social change in client projects. In E. L. Angeli & T. K. Fountain (Eds.), *Proceedings of SIGDOC'18* (Paper no. 17, pp. 1–6). New York, NY: ACM.

Lauer, C. (2009). Contending with terms: "Multimodal" and "multimedia" in the academic and public spheres. *Computers and Composition, 26*(4), 225–239.

Lawson, B. (1980). *How designers think: The design process demystified*. New York, NY: Routledge.

Leverenz, C. (2014). Design thinking and the wicked problem of teaching writing. *Computers and Composition, 33*, 1–12.

Lewis, B. (1999, April). Mediating texts in engineering education. Paper presented at the meeting of the American Educational Research Association, Montreal, Canada.

Marback, R. (2009). Embracing wicked problems: The turn to design in composition studies. *College Composition and Communication, 61*, 397–419.

Medway, P. (1996a). Writing, speaking, drawing: The distribution of meaning in architects' communication. In M. Sharples & T. van der Geest (Eds.), *The new writing environment* (pp. 25–42). London, UK: Springer.

Medway, P. (1996b). Virtual and material buildings: Construction and constructivism in architecture and writing. *Written Communication, 13*(4), 473–514.

Miller, C. R. (1985). Invention in technical and scientific discourse: A prospective review. In M. G. Moran & D. Journey (Eds.), *Research in technical communication: A bibliographical sourcebook* (pp. 117–162). Westport, CT: Greenwood.

Minton, A. (1941). Design for composition. *English Journal, 30*(2), 136–146.

Norman, D. (2013). *The design of everyday things*. Revised edition. New York, NY: Basic Books.

Norman, D., & Draper, S. W. (Eds.). (1986). *User-centered system design: New perspectives on human-computer interaction*. Hillsdale, NJ: Lawrence Erlbaum Associates/CRC Press.

Papert, S. (1980). *Mindstorms: Children, computers, and powerful ideas*. New York, NY: Basic Books.

Papert, S. (1993). *The children's machine*. New York, NY: Basic Books.

Pausch, R., & Zaslow, J. (2008). *The last lecture*. New York, NY: Hyperion.

Piaget, J. (1952). *The origins of intelligence in children*. New York, NY: International Universities Press.

Piaget, J. (1957). *Construction of reality in the child*. London, UK: Routledge.

Piaget, J. (1973). *To understand is to invent: The future of education*. New York, NY: Grossman.

Pope-Ruark, R., Tham, J., Moses, J., & Conner, T. (2019). Design thinking approaches in technical and professional communication. Special issue of *Journal of Business and Technical Communication*, 370–465.

Porter, J. (2002). Why technology matters to writing: A cyberwriter's tale. *Computers and Composition, 20*(4), 375–394.

Purdy, J. (2014). What can design thinking offer writing studies? *College Composition and Communication, 65*(4), 612–641.

Redish, J., & Barnum, C. (2011). Overlap, influence, and intertwining: The interplay of UX and technical communication. *Journal of Usability Studies, 6*(3), 90–101. Retrieved from https://uxpajournal.org/wp-content/uploads/sites/8/pdf/JUS_Redish_Barnum_May_2011.pdf

Rifenburg, M. (2014). Writing as embodied, college football plays as embodied: Extracurricular multimodal composing. *Composition Forum, 29.* Retrieved from http://compositionforum.com/issue/29/writing-as-embodied.php

Rittel, H., & Webber, M. (1973). Dilemmas in a general theory of planning. *Policy Sciences, 4*(2), 155–169.

Rouse, W. B. (2007). *People and organizations: Explorations of human-centered design.* Hoboken, NJ: John Wiley & Sons.

Rowe, P. (1987). *Design thinking.* Cambridge, MA: Massachusetts Institute of Technology Press.

Selting, B. R. (2002). Conversations with technical writing teachers: Defining a problem. *Technical Communication Quarterly, 11*(3), 251–266.

Sheridan, D. (2010). Fabricating consent: Three-dimensional objects as rhetorical compositions. *Computers and Composition, 27*(4), 249–265.

Shipka, J. (2005). A multimodal task-based framework for composing. *College Composition and Communication, 57*(2), 277–306.

Shipka, J. (2011). *Toward a composition made whole.* Pittsburgh, PA: University of Pittsburgh Press.

Spilka, R. (2002). Becoming a profession. In Barbara Mirel & Rachel Spilka (Eds.), *Reshaping technical communication: New directions and challenges for the 21st century* (pp. 97–110). Mahwah, NJ: Lawrence Erlbaum Associates.

Stanford d.school. (n.d.). Design thinking bootcamp bootleg. Retrieved from https://static1.squarespace.com/static/57c6b79629687fde090a0fdd/t/58890239db29d6cc6c3338f7/1485374014340/METHODCARDS-v3-slim.pdf

Stolley, K. (2008). The lo-fi manifesto. *Kairos, 12*(3). Retrieved from http://kairos.technorhetoric.net/12.3/topoi/stolley

Stolley, K. (2016). The lo-fi manifesto, v. 2.0. *Kairos, 20*(2). Retrieved from http://kairos.technorhetoric.net/20.2/inventio/stolley/index.html

Swan, K. (2005). A constructivist model for thinking about learning online. In J. Bourne & J. C. Moore (Eds.), *Elements of quality online education: Engaging communities* (pp. 13–30). Needham, MA: Sloan-C.

Swarts, J. (2020). Technical communication is a social medium. *Technical Communication Quarterly.* Online first edition. Retrieved from www.tandfonline.com/doi/abs/10.1080/10572252.2020.1774659?journalCode=htcq20

Vaughn, M. A. (2017). Why making matters: Pedagogy in practice. In *Proceedings of international symposium on academic makerspaces* (Paper no. 40). Cleveland, OH: ISAM.

Verganti, R. (2009). *Design driven innovation: Changing the rules of competition by radically innovating what things mean.* Boston, MA: Harvard Business School Publishing Corporation.

Verhulsdonck, G., Howard, T., & Tham, J. (Forthcoming). Converging streams: Implications of design thinking, content strategy, and artificial intelligence as developments in technical communication and user experience design. *Journal of Technical Writing and Communication*.

Vygotsky, L. (1978a). *Mind in society*. Cambridge, MA: Harvard University Press.

Vygotsky, L. (1978b). Interactions between learning and development. In Mary Gauvain & Michael Cole (Eds.), *Readings on the development of children* (pp. 34–40). New York, NY: Scientific American Books.

Weedon, S. (2019). The core of Kees Dorst's design thinking: A literature review. *Journal of Business and Technical Communication, 33*(4), 425–430.

Why you should be doing design challenges. (2018). Retrieved from www.justinmind. com/blog/why-you-should-be-doing-design-challenges/

Wible, S. (2020). Using design thinking to teach creative problem solving in writing courses. *College Composition and Communication, 71*(3), 399–425.

Winsor, D. A. (1996). *Writing like an engineer: A rhetorical education*. Mahwah, NJ: Lawrence Erlbaum Associates.

Yalanska, M. (2018). Human-centered vs user-centered. Are the terms different? *FAQ Design Platform*. Retrieved from https://tubikstudio.com/faq-design-platform-human-centered-vs-user-centered-are-the-terms-different/

Yancey, K. B., & Spooner, M. (1998). A single good mind: Collaboration, cooperation, and the writing self. *College Composition and Communication, 49*(1), 45–62.

Yergeau, M. (2015). Design. In P. Heilker & P. Vanderberg (Eds.), *Keywords in writing studies* (pp. 51–56). Logan, UT: University Press of Colorado/Utah State University Press.

Zachry, M., & Spyridakis, J. H. (2016). Human-centered design and the field of technical communication. *Journal of Technical Writing and Communication, 46*(4), 392–401.

Zhang, Y., & Kitalong, K. S. (2015). Influences on creativity in technical communication: Invention, motivation, and constraints. *Technical Communication Quarterly, 24*(3), 199–216.

2

THE MAKER MOVEMENT AND ITS INFLUENCES ON TECHNICAL COMMUNICATION AND HIGHER LEARNING

A Look at Three Makerspaces

Overview: This chapter begins with a synthesis of current discussions on the need for technical communication to turn to design-centric and material thinking, which prioritize innovative solutions that are user-oriented. This synthesis leads to the observation of recent developments in grassroot DIY culture and a fast-growing international movement called the Maker Movement. This chapter provides a survey of the history and technology of making that perpetuated the Maker Movement. It forwards an argument that making, design thinking, and problem solving are inherently intertwined and that technical communicators are increasingly finding themselves in situations that require such thinking and doing. Through an ethnographic study of three makerspaces, this chapter shows the impact of maker culture on higher education. It uses the findings from the makerspace ethnography study to reiterate the connections between making and design thinking and technical communication.

The Materiality of Technical Communication and Its Pedagogy

Like many kids, I was obsessed with LEGOs growing up. My siblings and I would spend hours building imaginary cities, cars, and household items like tables and cabinets using a variety of bricks and pieces gifted to us by relatives as well as hand-me-downs by the older children of my parents' friends. We used our imagination to construct things that complemented other entities of our make-believe play world—a cup for Teddy bear, a chair for Barbie Ken, a hat for the Red Power Ranger, among others. When playtime was over, our mom would always make us put away everything, including our proudest creation of the day. Despite the slight disappointment, my siblings and I would always do as told because we knew the LEGOs would still be there the next day and we'll get to build something new again.

Fast forward 20 odd years, I still find myself using LEGO bricks but in a rather different context. Now as a technical communication instructor, I teach units on instructional design where I introduce students—most of them aspiring technical writers, engineers, and business leaders—to the principles of effective technical instructions. For the purpose of demonstrating the process of creating usable instruction sets, I deploy a module I learned from the advanced writing instructors at the University of Minnesota (where I completed my PhD) that lets students experience the not-so-common-sensical process of putting together technical instructions and manuals. And yes, this module involves LEGOs.

When I first taught this module as a graduate instructor, I would gather bricks and pieces from anywhere I could get them and loan them to students. The students, working in teams of three to four, would create a unique construction and compose a set of instructions to an intended user for replicating the construction. The assignment usually starts at a relatively high point when the students learned they could use LEGOs in a course project. It ends with the celebration of some fine instruction sets but also distinguishable disappointments from the students. That disappointment is akin to what my siblings and I felt as kids at the end of playtime when we had to, albeit unwillingly, disassemble our prized creation and return them to the toy bins. Similarly, the students were unwilling but they had to undo their trophy LEGO models and return the bricks and pieces to me.

My students' response to using LEGOs helps demonstrate the material and affective dimensions of composition and communication. Whether grade schoolers or college students or industry practitioners, we all assign meanings and different levels of affection to both the process and product of our work. In the context of technical communication, most practitioners take creatorship and ownership seriously; they instill pride and a sense of identity in what they do. One will know what this means if they have ever experienced critiquing the works of a designer or writer. Rhetoric and writing scholar Christina Haas (1996) in *Writing Technology* has particularly pointed out the material dimensions of writing, with the term "material" referring to anything that possesses mass or matter, and which uses physical space. For the practice of technical communication, this includes any tools or resources that cross between the communicators and their artifacts. The material elements of the communicative space, the pencils, desks, chairs, screens, keyboards, and other communication materials also function as heuristics for learning. The connections between materials, creators and users, and the literacy knowledge in the communicative environment are often mapped onto the socio-material conditions of learning as a way of problematizing their relations to the wider societal issues.

In writing and communication studies, scholars have increasingly relied on activity and circulation theories to understand the mediating power of tools as tied to knowledge-making and dissemination (see Prior & Shipka, 2003;

Trimbur, 2000). These socio-cultural and historical approaches to composing and communicating emphasize the active and dynamic role of tangible materials, and the vitality of their interplay with learning. Former chair of the Conference on College Composition and Communication Kathleen Blake Yancey (2004) has also considered affects and emotions to be a part of the corporeal experience of composition:

> A composition is an expression of relationships—between parts and parts, between parts and whole, between the visual and the verbal, between text and context, between reader and composer, between what is intended and is unpacked, between hope and realization. And, ultimately, between human beings.
>
> —*Kathleen Blake Yancey (cited in Shipka, 2011, p. 9)*

The "thingness" of writing and communication asks us to pay attention to the roots of the material technologies used in our workplace or academic settings, and challenges us to understand critically the social and cultural influences of these technologies on our embodied practices. For example, technical communicators who work in user-experience design teams should consider the history of an essential technology—like the Adobe Creative Suite—by learning its past, the cultural factors that influenced the design of the tool, and the social and physical effects the tool imposes and on whom.

To this end, professional communication scholar Kelli Cargile Cook (2002) has presented a "layered literacies" theoretical framework for technical communication pedagogy. This layered approach integrates the social, rhetorical, and critical literacies with "basic" (technical), technological, and ethical literacies. Cargile Cook argued that these literacies should be taught in combination rather than isolation. Further, she pointed out the need for ongoing revision of our pedagogy to meet the evolving demands of the workplace: "workplace writers need a repertoire of complex and interrelated skills to be successful. Instructors can no longer simply provide students with opportunities to discuss form, discourse types, or the writing process. Such discussions must be further supplemented with activities that promote collaborative team-building skills and technology use and critique" (Cargile Cook, 2002, p. 8).

Given the expanding workplace literacy needs here at the advent of the digital age, the exigence prompting Cargile Cook's call for pedagogical improvements remains but to those changes I would add that it is even more important to focus on the materiality of technical communication practices and pedagogy beyond technological and technical competencies (e.g., knowing how to code, writing in single-source systems, designing technical diagrams, visualizing data, etc.). As education administrators and program directors continue to push instructors toward new technologies and learning platforms, technical communication scholars occupy an important space in helping administrators and instructors understand the material layer of these tools. In leveraging their rhetorical and technical expertise, our scholar-teachers could recognize and examine emergent strategies involving new technologies for the enhancement of teaching and learning.

An excellent argument toward this material direction in the context of technical communication pedagogy was forwarded by Jennifer Bay, Richard Johnson-Sheehan, and Devon Cook (2018), who encouraged program administrators to introduce principles and methods of design thinking—as introduced in the previous chapter—in technical communication courses so students could practice applying the technical communication concepts in real-world entrepreneurial situations. These professional writing scholars argue that design thinking is suitable for introducing students to concepts such as audience, user-centered design, usability, collaboration, and leadership. More importantly, design thinking helps reinvigorate departments such as English, Rhetoric, and Communications that are increasingly challenged to respond to rapid changes in the economy and our students' career interests. They maintained,

> Technical communication must evolve to meet these new challenges. We must teach our students how to have empathy for users, peers, and stakeholders, just as we must have empathy for the needs of our students. We must define educational problems from our students' points of view, not our own, and we need to ideate those problems by reframing them and incorporating new technology. We need to prototype new assignments and new activities and then do testing to see which ones work.
>
> (Bay et al., 2018, p. 193)

Indeed, a turn to materiality means attuning to a design-centric pedagogy that aligns technical communication primarily as a problem-solving activity. This alignment augments the recent call for technical communicators to assume leadership in seeking out, devising, and actualizing cross-sector solutions to complex global problems. More importantly, a turn to materiality, as Bay and colleagues outlined, emphasizes a more tangible "doing" in the learning process than mere conceptualization. Thanks to the design thinking framework, this "doing" can be translated into experimentation with materials and tools, tinkering with unfamiliar technologies, and building impactful solutions to existing problems.

Many may have already recognized these aforementioned exercises to be the core characteristics of the recently resurged DIY (do-it-yourself) culture, fondly known as the Maker Movement. As I elaborate later, "making" as an experiential, problem-based learning component can be a strategic direction for technical communication pedagogy. With the exigence I presented in the introduction chapter that situates technical communicators—practitioners and instructors alike—as design leaders and changemakers, I regard the Maker Movement to be an important touchstone for our collective examination and learning. To make my case that the making should be a revitalizing force in technical communication pedagogy, I provide next a history of the Maker Movement—how it came to be and where it's taking us.

The Maker Movement: An Industrialist Legacy

The Maker Movement is an informal, umbrella term referring to an emerging subculture arising from grassroots networks through a shared interest in collective or collaborative tinkering on creative and technical projects. Broadly, the Maker Movement is propelled by a culture of making and "hacking" that favors democratic and meritocratic conventions to organized production. It encourages bottom-up organizational practices that seek to foster open and social learning. The culture of making is typically associated with design thinking as its methodology guides human-centered solutions and iterative processes (Australian Institute for Teaching and School Leadership, 2014; Sayers, 2017). Dale Dougherty, founder of *Make:* magazine—one of the core media outlets promoting the Maker Movement—and creator of Maker Faire, described the Maker Movement as follows,

> When I talk about the maker movement, I make an effort to stay away from the word "inventor"—most people just don't identify themselves that way. "Maker," on the other hand, describes each one of us, no matter how we live our lives or what our goals might be.
>
> *(Dougherty, 2012, p. 11)*

On a more tangible level, the Maker Movement can be recognized by the rapid-prototyping tools and methods it employs. *Make:* magazine defines the Maker Movement as a "combination of ingenious makers and innovative technologies such as the Arduino microcontroller and personal 3D printing [that] are driving innovation in manufacturing, engineering, industrial design, hardware technology and education" (n.d.). Besides the technologies that support making, the Maker Movement can also be characterized by collective organizations that maintain workshops for projects involving welding, metal or woodwork, and electronic circuit design and programming. These physical spaces are fondly known as digital fabrication labs (or fablabs), techshops, hackerspaces, or more generally makerspaces. Each of these space classifications has their own unique emphasis.

Fablabs are popularized by the Stanford University Graduate School of Education's FabLearn program and the MIT's Fab Foundation (https://fabfoundation.org/). Fablabs emphasize learning through research and invention. Techshops, popularized by the enterprise chain TechShop, are typically membership-based community studios equipped with industrial tools for fabrication. While fablabs and techshops are more structured by their US-based organizational philosophies, hackerspaces trace their origin to the European hacker culture. Hackerspace members usually tell a story of when their "founding leaders," a group of

North American computer programmers visited Germany's Chaos Computer Club in 1997 (Maxigas, 2012) and grew excited about creating similar spaces in the US. Semiotic disputes over the terms hackers and hacking have not stopped hackerspace users to stand firm on their theoretical perspective. Today, hacking typically refers to creative ways to work around everyday life issues. Terms like "lifehacks," "schoolhacks," and "gradhacks" (specific to graduate students) were grown out of this tradition.

Makerspace, however, tends to describe an open workspace dedicated to maker culture practices. Makerspace is a term coined by *Make:* magazine when it was launched in 2005. It became further popularized when Dougherty registered makerspace.com in 2011 and started using the term to refer to open-access spaces for designing and creating (Cavalcanti, 2013). Within academic settings, schools, libraries, and universities tend to call their design and production spaces makerspaces given the neutrality in the name. Educause (2013) identifies a makerspace as "a physical location where people gather to share resources and knowledge, work on projects, network, and build" (n.p.). One of the most popular academic makerspaces, the Vassar College makerspace, defined their facilities as spaces that

> combine manufacturing equipment, community, and education for the purposes of enabling community members to design, prototype and create manufactured works that wouldn't be possible to create with the resources available to individuals working alone. These spaces can take the form of loosely-organized individuals sharing space and tools, for-profit companies, non-profit corporations, organizations affiliated with or hosted within schools, universities or libraries, and more.
>
> *("What is a makerspace," 2015, n.p.)*

The term *academic makerspaces* is increasingly used to specify a distinguishable field dedicated to studying makerspaces in higher education contexts.

The influence of the Maker Movement in the academy is evident in the growing development of makerspaces in colleges and universities across the United States. To share resources and address makerspace-related problems collaboratively, leading institutions including Yale University, Stanford University, Carnegie Mellon University, MIT, and others have joined forces to form the Higher Education Makerspaces Initiative (HEMI). Since 2016, HEMI has been responsible for convening the annual International Symposium on Academic Makerspaces (ISAM), a conference that brings together makerspace managers and researchers from around the world to identify and address emerging issues around academic makerspaces. These issues span from technical design (i.e., how to track user traffic in a makerspace) to pedagogical implications (i.e., what kind of curriculum might be created around making).

Certainly, the Maker Movement did not just begin a few years ago; its roots are connected to industrialism and mass manufacture. According to historians

Elizabeth Cumming and Wendy Kaplan (1991), designers and labor theorists in Victorian England have created an early reaction to industrialization. They have sought to value individualism and creativity amid a time of profits and mass-market capitalism. New media scholar Chet Breaux (2017) noted that art critic John Ruskin was another important figure in the movement that publicly called for organic design and production and the end of the machine-driven model of Victorian production. By the 1890s, as Cumming and Kaplan documented, there had been several large craft shows that occurred throughout England, marking the golden age of the Arts and Crafts Movement.

In their book, *Adhocism*, architects Charles Jencks and Nathan Silver (2013) reported that the Arts and Crafts Movement suffered a visible decline following the First World War. This has led to the rise of the ad hoc DIY practice in the 1920s, a new method of assembling using readily available components and tools. Jencks and Silver suggested that doing-it-yourself is "the rebirth of a democratic mode and style, where everyone can create his personal environment out of impersonal subsystems, whether they are new or old, modern or antique. By realizing his immediate needs, by combining ad hoc parts, the individual creates, sustains and transcends himself" (p. 15). For Jencks and Silver, this form of creative and self-powered assemblage is a way of resisting the "omnipresent delays caused by specialization and bureaucracy" (p. 19). It resembles a postmodern viewing of a pluralist world containing multiple ideologies—fragmented, but can be reassembled as necessary, yet not always cohesive. Jencks and Silver also pointed to the counterculture movement coinciding with industrial and cultural forces as the roots of DIY culture. Particularly, the emphasis on reusing or repurposing industrial excess serves a great example of adhocism at work (pp. 65–67).

In observing the effects of adhocism, American writer Evgeny Morozov (2014) noted in *The New Yorker* that although the Arts and Crafts Movement was deemed dead after the First World War, the sentiment behind it lingered. "It resurfaced in the counterculture of the nineteen-sixties, with its celebration of simplicity, its back-to-the-land sloganeering, and, especially, its endorsement of savvy consumerism as a form of political activism," Morozov (2014, n.p.) wrote. Evidently, it wasn't just for political purposes but business marketing as well. Morozov highlighted the corporate gimmicks organizations like Apple, Inc., Stanford's Artificial Intelligence Laboratory, and even MIT used to convince consumers that they were rebels. The hippie term "hackers" that stemmed from the European hacker culture became a slang for those who distinguished themselves from the rigid, unimaginative technocrats. Soon, the talk of "de-institutionalization of the society" with rising personal computing technologies became a slogan for self-proclaimed anticulture tech elites, many of whom were also subscribers of Stewart Brand's *Whole Earth Catalog*, an American counterculture magazine that promoted self-sufficiency, DIY, and holism, circa 1968–1972 (Morozov, 2014). Brand's counter-mainstream rhetoric is deeply ingrained into the culture of making today.

These historic movements and influences are important to establish the lineage of making. While maker practices did not emerge overnight, many developments and continued ideologies of crafting, self-assembling, and hacking demonstrate the persistence of these ideas. The Maker Movement traces its lineage to a tradition of artisanship, self-sufficiency, and the subsequent anticulture techno-enthusiasm. What differs the Maker Movement from its preceding history, however, is the infrastructure that allows makers to become truly makers—the well-equipped makerspaces and community of practice that celebrate DIY mindset and entrepreneurship. In the next section, I discuss the impact of these infrastructural elements—and culture—that fuel the Maker Movement.

While there were no specific events that led to the booming of the maker culture, the notion of making as an intentional, inventive, and innovative practice is popularized by narratives around emerging technological solutions and rapid prototyping as they are increasingly supported by affordable desktop manufacturing technologies like 3D printers. A common belief about the maker culture is that they are a computer-based, technology-enhanced extension of the DIY culture. In his book, *The Maker Movement Manifesto*, Mark Hatch (2013) describes how American culture perpetuates the maker culture:

> Wars have been fought when the common people thought they were going to lose access to ownership of their own productive tools. So the first thing we must do is make. The do-it-yourself (DIY) home improvement industry in the United States is worth over $700 billion. The hobbyist segment is worth over $25 billion. The most valuable segment of the $700 billion DIY is the perpetual remodeler, specifically those who have enough money to let business professionals do the work for them, but don't. You might know or even be one of these people. In your heart of hearts, you know you don't really need to redo the bathroom, or certainly not the way you plan to do it, yourself. But you do it anyway. This is because there is more satisfaction in completing the remodel yourself.
>
> *(pp. 12–13)*

Besides the "satisfaction" factor, schools and homes have continued to encourage making as a productive and desirable endeavor. The capitalist society has slowly moved from valuing originality to applauding different means of expression that involve modification, remix, and redistribution. In schools, students across all education levels are taught to discuss how they feel about the texts they encounter. They are usually asked to respond by composing syntheses of texts with their own reflections. At home, children are taught to build and fix. Parents give young children toys like those of my anecdote at the start of this chapter to encourage imaginative building. When they are older, teenagers are taught household maintenance, such as changing a lightbulb, fixing a leak, and building a shelf. These activities often add to or modify the existing design based on the purposes or constraints the makers are working with. Generally speaking, this modern culture subscribes to a belief in taking matters into one's own hands—solving problems on their own. Such a culture, plus an increasingly affordable access to additive manufacturing technologies and fabrication tools, propels the maker culture in formal and informal education. To this end, literacy researcher Mike Rose (2014) drew a connection between the maker culture and the education systems:

> We seem to have discovered the pleasures of working with our hands—or at least of using products that are handmade or manufactured on a small scale, artisanal, locally produced. . . . In education, there is growing interest in making and "tinkering" to foster, in one organization's words, "imagination, play, creativity, and learning." As opposed to some anti-technology expressions of this hands-on spirit in the modern West, our era's movement embraces technology—computers and digital media are as much a part of the Makers Movement as woodworking and quilting. The same holds for education, which wants to draw on young people's involvement in computer technology and social media.
>
> *(2014, n.p.)*

Per Rose's observations, American education is not only already submerged in the maker culture but also in nurturing makers. In the context of the Maker Movement, a maker is a blanket name for creators, designers, developers, programmers, etc.—all those who go beyond just thinking about ideas into tinkering with different ways to materialize their ideas. Several characterizations[1] of makers set them apart from any creator. Makers embrace an entrepreneurial spirit that motivates them to pursue radical solutions and are biased toward actions. While they do not necessarily have to exert high energy at all times, makers are often passionate about their ideas and that passion is reflected in their designed artifacts. Since collective work is a signature characteristic of the Maker Movement, makers often engage in sharing (ideas, tools, spaces) and collaborating with others.

When makers participate in shared events and collective invention, they form a network called a maker community. A common maker community is Make: Projects (https://makeprojects.com/home), an online project space alongside Dougherty's enterprise where makers share ideas, methods, tools, and directions for perfecting one another's projects. Maker communities also manifest in the form of an in-person project showcase, called the Maker Faire. Maker Faires are locally organized events (similar to TEDx talks) where cities or counties work with the chief sponsor *Make:* magazine and local makerspaces to put together a series of showcases and competitions. According to the official Maker Faire website, these events are "an all-ages gathering of tech enthusiasts, crafters, educators, tinkerers, hobbyists, engineers, science clubs, authors, artists, students, and commercial exhibitors. All of these 'makers' come to Maker Faire to show what they have made and to share what they have learned" ("Maker Faire: A bit of history," n.d.). In the Twin Cities, for instance, the annual Minneapolis-St. Paul Mini Maker Faire has been held every summer at the Minnesota State Fairgrounds between 2014 and 2018. Makers or exhibitors can host a booth, give a presentation, lead a workshop or be a performer at the Maker Faire. In 2017, The Minneapolis-St. Paul faire featured an Education Day for 7th graders and teachers to try their hands at coding, flying a drone, screen printing, soldering and building their own projects. The Education Day was featured again in conjunction with the 2018 Maker Faire plus 1,000 free spots for students in schools that do not meet the $5-per-student threshold.

Making in Academic Settings: A Study of Three Makerspaces

Circling back to the attunement to a design-centric pedagogy for technical communication, I recognized the value of making as an educational exercise with great potential for rejuvenating our teaching and learning (Tham, 2020). When teaching as a graduate instructor at the University of Minnesota, I saw making as a pedagogically sound activity that could introduce students to many core concepts of technical communication and give them an opportunity to identify and address ambiguous problems. Hence, I devoted my doctoral research to an investigation of the viability for cultivating a maker culture in the technical communication classroom. I began my study by looking outward to other disciplines in which making is already ingrained in their curricula. Upon consultations with local makers and technology enthusiasts from my network, I have come to learn about the growing network of academic makerspaces in the United States. HEMI, the Higher Education Makerspace Initiative, has gained steady traction among leading institutions with strong engineering and design programs. The Fab Foundation, an international non-profit organization that emerged from MIT's Center for Bits & Atoms to serve as a networking resource for digital fabrication communities, has reached members and partners from over 90 countries around the world.

Despite these robust networks, technical communication has virtually zero presence in them. The closest affiliation I could find after hours of keyword search and scanning of membership data was in English educators—primarily in K-12 settings—who were interested in shaking up their conventional language instruction by introducing maker activities. I also found Angela Stockman's (2016) book, *Make Writing: 5 Teaching Strategies that Turn Writer's Workshop into a Maker Space*, to be a popular guide for these literacy educators who sought to connect making and writing. Within the rhetoric and composition community, some scholar-teachers have integrated making with their writing classrooms and published compelling findings about its benefits (note especially Breaux, 2017; Craig, 2014; Elam-Handloff, 2016; Brown & Rivers, 2013; Sheridan, 2010). However, like Stockman, most of these studies discussed making in the contexts of composition—many within first-year writing—instead of the technical communication proper. I considered this a missed opportunity for our field.

To make the case for making in technical communication pedagogy, our field needs a better understanding of academic makerspace cultures beyond our occasional examinations of 3D printing (Roy, 2016) and interactions with community makers. This understanding could be gained by observing academic makerspaces *in-situ*—witnessing their physical structures and setups, learning about their workflow and operational process, and speaking to student makers. For this reason, I selected three academic makerspaces for in-depth site visits. They were the Anderson Student Innovation Labs at the University of Minnesota, the Invention Studio at Georgia Tech, and the Think[box] at Case Western Reserve University. These makerspaces were among a handful of options first identified from Andrew Barrett and colleagues' (2015) review of academic makerspaces in the United States. These sites were selected due to their nature as a university makerspace—not just an engineer's shop—as well as their establishment in the North American academic makerspace community. All three of the selected sites were regarded by their peer institutions as exemplary models.

Due to the nature of my study and funding, I was able to spend on average two days at each site performing ethnographic observations of the makerspace, the interactions between makers and the space, and overall maker culture through the respective workflow and administration. In the following sections, I begin with a brief introduction of each makerspace with their history, followed by my findings on their respective setup, workflow and processes, and maker experience on each site. I sum up my observation with a comparative analysis of the three makerspaces by highlighting their common features and unique elements. To protect their privacy, all student names have been replaced with pseudonyms.

The Anderson Student Innovation Labs at University of Minnesota—Twin Cities (UMN)

I began my journey by visiting the Clifford I. and Nancy C. Anderson Student Innovation Labs (also known as the Anderson Labs; https://cse.umn.edu/andersonlabs) from my home base in Minneapolis. This makerspace was what I would call a makerspace conglomerate; it's made up of three separate labs—Student Design Lab, Student Shop, and Student Machine Shop—all supported by the UMN College of Science and Engineering (CSE) at the time of my study. The 10,000 square feet facility (all three labs combined) was initially home to several wood and metal shops where engineering students practiced wood and metalworking, welding, milling, and electronic circuitry. It was reimagined as a makerspace in 2016, after receiving a generous donation from Clifford and Nancy Anderson, with the addition of two new design and prototyping labs, and a major upgrade to an existing shop. The goal of this revitalized space was to focus on experiential learning and help students turn their design into reality.

I was introduced to the Anderson Labs by Jonathan Koffel, a health sciences librarian turned emerging technology and innovation strategist at Minnesota. From our initial meetings, I was put in contact with William Durfee, head of the mechanical engineering department and faculty sponsor for the Anderson Labs. Durfee then introduced me to two very important individuals. The first was Ben Guengerich, manager of Anderson Labs. Guengerich was a key informant who provided me with tours and detailed explanations of the functions of the Anderson Labs. The second individual Durfee introduced me to was Josh Halverson, then-senior mechanical engineering student who was completing an honors thesis examining the impact of academic makerspaces on engineering students. Halverson provided me with insightful perspectives on the uses of makerspaces from a student's point of view.

The Anderson Labs Setup

The three separate Anderson Labs spaces were not located in the same building. The Student Design Lab and Student Machine Shop were in the Mechanical Engineering building, and the Student Shop was placed in the Civil Engineering building, all of which were on the East Bank campus of UMN. There were underground tunnels that connected them. The official reception of the Anderson Labs was the Student Design Lab on the second floor of the Mechanical Engineering building.

The Student Design Lab was a large workspace equipped with workbenches, tables, hand tools, some power tools, laser cutters, computers, and 3D printers. According to Guengerich, the primary purpose of this lab was to allow students to test out their design through rapid prototyping and modeling. It has open meeting pods with chairs and whiteboards that let students collaborate or discuss ideas. The Student Design Lab was open seven days a week during the regular semester. For welding and more intensive woodworking, students would need to use the Student Shop in the Civil Engineering building. This lab was a half-open workspace with 3D scanners, 3D printers, materials testing load frame, and woodwork facilities. The Student Shop was also open seven days a week during the semester. If students wanted to perform metalworks, they would need to use the Student Machine Shop in the Mechanical Engineering building. The machine shop was staffed by professional machinists with metalworking mills, lathes and grinders, milling machines, and waterjet cutters. Given the staffing hours, this lab was open only Monday through Friday during the regular semester.

Access to Anderson Labs

Access to the Anderson Labs was granted to students enrolled in an engineering/CSE course or in any of the CSE programs. Non-CSE faculty and students could access the makerspace if they were collaborators on a CSE-related project or had received permission from Guengerich, the lab manager. For instance, when my department and I organized the 2017 Great Plains Alliance for Computers and Writing, participants of our pre-conference workshop on smart material technologies were allowed to use the Student Design Lab because one of our collaborators, Dr. Julianna Abel, was a CSE/mechanical engineering faculty. Later in my interview with another CSE faculty, I learned that some non-CSE faculty had also been granted access to the labs in the past because their research contributed to the greater CSE missions.

Maker Experience at Anderson Labs

To understand the culture of making in the makerspace beyond its spatial setup and administrative workflow, I conducted semi-structured interviews with students who were using the labs during my visit:

- How long have you been using this makerspace?
- How often do you use this makerspace?
- Who/what introduced you to this makerspace?
- What do you use this makerspace for?
- When do you prefer to come to this makerspace?
- What project(s) are you working on? Tell me more about it/them.
- Is/are your project(s) connected to any class you are enrolled in? If so, in what ways?
- How does this makerspace support your project(s)?
- Do you think you might be able to complete your project elsewhere or outside this makerspace? Why?
- What is your overall impression about this makerspace?
- What do you think of the resources and the staff members of this makerspace?
- How might this makerspace better serve your needs?
- Would you recommend this makerspace to others? To whom and why?
- Any additional comments?

According to a mechanical engineering student, Adam, who identified as a frequent user of the Anderson Labs, the makerspace was a response to the growing need for fabrication equipment for students. When asked about his evaluation of the makerspace, Adam shared the observation that students appreciated having dedicated workbenches and fabrication equipment. They were able to "see each other's progress, share advice, and occasionally receive ad hoc mentorship from older students," Adam said. Adam also emphasized how students groups have made the Anderson Labs their meeting place:

> Student organizations immediately found a natural home in the new space. Their educational workshops have been able to accommodate a larger

number of students because of access to tables and tools in a permanent and spacious location. The Anderson Labs have legitimized their freedom to create learning experiences, more powerful than those in the classroom because they are founded in camaraderie and peer mentorship.

Another student interviewee, Mickey, was a student worker at the Anderson Labs responsible for maintaining equipment and assisting makers. When I asked about his experience with working there, he noted how the space is difficult to find on campus and therefore not getting many visits from students. He especially regarded the underground tunnels to be confusing even for CSE students. Mickey also observed that, based on his one-year experience working at the labs, there were more male students than female students using the makerspace. He pointed out how that could be a problem for the growth of the makerspace:

> I think the lab will present cultural barriers to new students who might use it. It is housed in the mechanical engineering building, a program with a historically low percentage of female students. Since mechanical engineering courses were early users of the space, it was not uncommon for me to walk into the lab and see 10 men using the space and not a single woman. I think it will be important to actively promote an environment where first-time users, regardless of their gender or familiarity with making, feel comfortable being in the space.

In fact, Adam had also shared the same empathy for students who aren't granted access to the labs. He believed that it defeats the purpose of a makerspace as a cross-disciplinary learning commons if access was only granted to engineering students:

> The fact that access to the Anderson Labs is limited to engineering students was incredibly frustrating to the student leaders who envisioned the space. Few classrooms allow students to work with peers studying a different major. The diversity of ideas and interests different majors would bring to the makerspace community is more than worth the meager cost to the College of Science and Engineering.

I agreed with the observations by these students—the physical location of the Anderson Labs was really its biggest hindrance to many potential makers. The labs were isolated from where students would typically meet and work (e.g., libraries, computer labs, student unions). As pointed out by Mickey, the lab in the civil engineering building was particularly hard to locate. For infrequent visitors, the labs' locations might be the biggest reason for makers to avoid the makerspace.

During my site visit, I also observed that makers performed more manufacturing work than design activities when using the Anderson Labs. The way the labs were set up encouraged students to cut, drill, and solder away their project rather than focusing on 2D design. The lack of computers and spaces for sketching, drawing, and modeling made it seem as though digital modeling and wireframing were not as important as actual building of a physical prototype.

The Invention Studio at Georgia Institute of Technology

My second site observation happened over the hot summer of 2017. I flew into Atlanta mid-July, got myself an Airbnb, and paid a visit to the Invention Studio at Georgia Tech (https://inventionstudio.gatech.edu). The Invention Studio was a 4,500-square-foot facility housed in the Manufacturing Related Disciplines Complex near the border of the campus, administered by the George W. Woodruff School of Engineering. The makerspace was founded in 2009, and has evolved over the years based on student and faculty use of the space. According to its history, the studio has always been supported by student volunteers. The Invention Studio took pride in being a fully student-run makerspace, a model followed by emerging makerspaces around the country and the world.

The Invention Studio was a useful site for my study because of its prestige and popularity in the academic makerspace community. Although I did not know of anyone from the Invention Studio prior to my visit, I was introduced to the makerspace's faculty sponsor by Guengerich from the Anderson Labs. As part of negotiating the IRB parameters for this study, there were many exchanges between me, the Invention Studio faculty sponsor, and one of their research faculty. After I was approved for the visit, I was referred to a student guide who was also a board member of the Invention Studio's official student organization.

The Invention Studio Setup

The Invention Studio was undergoing an upgrade at the time of my observation. The makerspace was under an expansion to occupy more space on the level of the building where it resided. Old offices were being removed to make room for the makerspace. During my visit, the makerspace was made up of four rooms—wood room, metal room, 3D printers and electronics room, and waterjet and laser room. By the end of the remodeling, the Invention Studio would combine some of these rooms to streamline the making process.

Because the entire facility was on the same floor, there was a distinctive "home" feel to the makerspace. At any given point of my visit, I could see 20–30 students roaming around the level and congregating in small groups around any open spaces they could find on the level. They seemed to have really made the makerspace their home; they left their belongings (backpacks, computers, food) around and would enter or leave the facility without needing any access cards or codes. In the hallways, I could hear radio music mixed with noises made by hand tools and printers.

Workflow and Processes at Invention Studio

I was informed by my student guide that the Invention Studio was open to anyone at Georgia Tech. During the regular hours, anyone with a university ID could access the rooms at the studio without obtaining prior permission. There were monitors set up as check-in stations in each of the rooms alongside with a card reader. Anyone entering each room should first swipe their ID at the station, and do so again when they are done with the room for the day. This process was meant for documenting traffic, rather than controlling access, the student guide told me.

In every room there was at least one Prototype Instructor and a Prototype Master. I have learned that Prototype Instructors were students who had undergone a specific certification program to be qualified as tutors to makers in the Invention Studio. Prototype Instructors were identified with a yellow armband while they worked in the makerspace. Prototype Masters were those who have been certified as a Prototype Instructor, and took additional training to become proficient in a specific power tool. They were identified with a red armband. There must always be a Prototype Master in the room for any makers to use a power tool.

During my visit, I did not see any faculty or staff members around the Invention Studio. All of the students and volunteers in the makerspace appeared to be very conscious about safety measures when using tools. When I was entering the wood room for the first time, I was immediately stopped by a nearby student (who was working on his project) and asked to put on a pair of safety glasses before proceeding. I then noticed that everyone in the room was already wearing their safety glasses, even if they were just working on their computer there. The makers also seemed diligent when using power tools. Some tools were marked as "training required" and that makers must ask for supervision from a Prototype Instructor.

It was revealed to me that since Prototype Instructors were not paid for their time or service, they would receive after-hours privilege access to the Invention Studio. Prototype Instructors and Prototype Masters could use the facility 24 hours a day as long as they were never alone in the makerspace.

Maker Experience at Invention Studio

To gather authentic maker experiences, I again performed interviews with the students who were there during my site visit. According to a student manager of the Invention Studio, the goal of the makerspace was to support student projects, whether they were class-related or personal. From a Prototype Instructor, Brian, I learned that volunteers strive to make the Invention Studio as welcoming a workspace as possible. They believe that no one would claim expertise in any project so everyone upholds an open mind and helps one another whenever necessary. Indeed, students were seen working in pairs or teams; in the two days I observed the studio, I rarely saw anyone sitting by themselves unless they were using a power tool. Brian added,

> Students tend to help one another with machines and ways of construct-ing something. . . especially those who are more experienced in the studio helping new users., like, "Oh hey, there is a better way to do that."

As a seasoned Prototype Instructor, Brain noted that students found the Invention Studio to be a home for their project even if they don't have a specific design in mind. The makerspace also functioned as a communal space for students:

> It is just a really nice community of people even if you're not working on a specific project. We are really trying to promote a maker culture to get stu-dents to work on hands-on projects. There are not a lot of classes at school that will give you the tools and resources to do that.

Teresa, a student user and board member of the Invention Studio, mentioned how important of a role the makerspace played for her decision to attend the university. She recalled visiting the Invention Studio during student recruitment events and the makerspace was the biggest reason she chose to attend Georgia Tech. She got involved with the makerspace early and signed up to become a Prototype Instructor even just in her first year.

Teresa also emphasized that peer mentorship was a core characteristic of the makerspace. She recalled how she learned from other users of the Invention Stu-dio who were not Prototype Instructors:

> A lot of the time the users are helping other users because they have a lot of experience with the machines and that other PIs were busy at the time. They help one another when they see that somebody looks like they have a question or they're unsure of how to use a certain machine. So there is a lot of collaboration going on even if the users are working on individual projects.

To promote innovation, the Invention Studio provided funding for Prototype Instructors to create innovative projects. According to Brian, there was an incen-tive program called The Maker Grant for PIs. Brian noted that the grant would encourage Prototype Instructors to enrich themselves through fun projects that included learning a new skill or two.

Lastly, a few of the makers chimed in when I was speaking with Brian and Teresa. From these students I learned that being a fully student-run makerspace, the Invention Studio gave its student board members autonomy in deciding the vision and identity of the makerspace. The students also mentioned they were able to decide, through consensus, the equipment or materials to purchase for the makerspace. The faculty sponsor served an advisory role to the makerspace student board.

Based on my visit, I realized the importance of ownership by makers in the makerspace. I noticed that students at the Invention Studio really turned the space into their own while still maintaining professionalism to ensure safety. Through my interviews and conversations with makers, I sensed a strong agency in the students; they declared a great deal of control of the makerspace and turned it into a community beyond academic purposes. On the practical side of things, I also noted the importance of flow in the makerspace setup. Compared to the Anderson Labs experience, the Invention Studio provided a much seamless experience as all the rooms were located in the same building and on the same floor. There was a sense of unity and easy access to tools and materials. Students used every corner of the floor to their own advantage, including a mini meal area where a public microwave sat. During my visit, I also saw some makers working in a team on an electric circuit project for a competition (they told me about it). They spread their tools and stuff across a bench in the common hub area and did not seem intimidated by passers-by. This was a very encouraging scene to me as a maker-enthusiast seeing those students taking advantage of a space afforded that kind of authentic making.

The Sears Think[box] at Case Western Reserve University

Having completed two successful site visits, my hopes were high for the last makerspace on my observation list. I was introduced to the nationally renowned makerspace at Case Western Reserve University (Cleveland, Ohio), the Larry Sears and Sally Zlotnick Sears Think[box] (https://engineering.case.edu/sears-thinkbox), by Guengerich at UMN. At the time when I was still finalizing my observation sites from the available academic makerspaces, Guengerich recommended Think[box] as a model makerspace given its esteemed staff members and the growing influence it had on academic makerspaces around the country. When I visited the makerspace in fall of 2017, Think[box] was hosting the second International Symposium on Academic Makerspaces (ISAM), having taken the baton from MIT, the previous symposium leader.

Among the most distinguishable features of Think[box] compared to its counterparts was its size. Think[box] was a 7-story, 50,000-square-foot facility—a standalone building dedicated entirely to being a full-scale makerspace to the Case Western campus. Think[box] began in December 2012 in a smaller, 5,000 square-foot space where protocols, training, and processes were tested that would be appropriate for an open-access mission. In October 2015, it moved into the first phase (Floors 1 to 4) of its permanent home, with renovations continuing and phase two completion of additional floors in Fall 2016.

The Think[box] Setup

Think[box] was open to all Case Western students, faculty, staff, and the Cleveland community at large. The makerspace branded itself as a center for entrepreneurship and innovation. The design of the building mirrored a 7-step process to a start-up business:

* Floor 1 — Community: a welcome center; gathering space
* Floor 2 — Collaboration: a meeting space to brainstorm ideas; collaborative ideation
* Floor 3 — Prototyping: the initial makerspace; digital prototyping and development
* Floor 4 — Fabrication: the next makerspace; non-digital construction and manufacturing
* Floor 5 — Project Space: a large space for teams to test their physical prototypes
* Floor 6 — Entrepreneurship: temporary cells for teams to assemble initial business endeavors
* Floor 7 — Incubator: temporary office spaces for startups

During my visit, I was allowed access to Floors 1, 3, 4, and 5. Students would typically occupy Floors 2 to 5, using the fabrication materials and tools to build their own projects. The layouts of the two main "making" spaces, Floors 3 and 4, were well defined and organized. The spaces were clearly marked with dedicated areas for reception, computer-assisted design or work, hand tools, power tools, hardware and materials, material disposals, electronics, higher-risk activities such as laser or waterjet cutting, and a "dirt room" where prototypes get sanded or spray painted (safety glasses required in this area). According to my tour guide, Think[box] was open about 63 hours each week in the regular semester, and about 20 percent of the traffic was from the public (non-Case Western community).

The Think[box] Workflow

As an open-access makerspace, Think[box] also served neighboring higher education institutions, nonprofit organizations, and industry around the area. Anyone walking into Floor 3, the main reception for the makerspace, would need to check in using a tablet at the reception desk. Makers were expected to familiarize themselves with safety measures and acquire knowledge of the power tools they plan to use on Floor 4. Unlike the Invention Studio, Think[box] did not have peer instructors who monitor the makerspace. It had student workers who are paid to assist makers with various tools. Think[box] received material donations from area industry, including plywood, filaments for 3D printers, and other raw materials. Makers did not need to pay for using these supplies in the makerspace.

On Floors 3 and 4, there were recycling and waste disposal spaces clearly marked to encourage makers to put away their unused materials.

The overall atmosphere at the makerspace seemed light and conducive for work. There was no background music. The student workers in the space were identified with their green apron and nametag. They walked around the space and were seen clearing up clutters and putting tools back into their storage areas. During my visit, there were 11 students in the makerspace on Floor 3. I observed one female student worker helping three students at the computer area at the same time. Another male student worker was cleaning a workbench, before he turned his attention to a maker who was trying to laser cut a wooden gift for his friends. The student worker offered to help the student, who turned out to be his classmate, to remove the stains on the wood after being cut up by laser, and he reminded his friend to wear goggles and gloves before entering the dirt room.

To access Floor 4, students must complete a few basic trainings online or one-on-one sessions with a staff. They would receive an "ability badge" after completion; and they must wear the badge when entering Floor 4. Students were also required to wear closed-toe shoes when visiting that floor. In case they forgot, there were safety clogs for checkout. According to a student worker, there were workshops, events, informal courses—anything that support, in their words, "an innovation ecosystem" (the student's words) of the makerspace. When I toured the first and second floor, I was shown many collaboration spaces that resembled an active learning classroom one may find in modern colleges or universities—with pods and monitors and whiteboards where makers could perform focused brainstorming and discussions. There was also free coffee and tea. Since the building was a little bit away from the main student center on campus, there was even a pizza vending machine on the first floor of Think[box].

My student guide informed me that to ensure student safety after business hours, a buddy system was enforced after hours. Makers must be in pairs in order to remain in the facility, and no one was allowed in the building after midnight.

Maker Experience at Think[box]

Similar to my first two visits, I spoke with students who were present at Think[box] during my observation. My first informant here was Nicola, who worked as a student staff at the makerspace. When asked about her overall experience, Nicola said she liked how the Think[box] configuration encouraged conversations among makers. The layout of the space seemed to support peer-to-peer feedback on projects. In addition, Nicola also highlighted how the staff members of Think[box] made her feel welcomed at the space, which encouraged her to visit Think[box] often. Sounding a little like a sales pitch, Nicola shared that both staff members and community volunteers (working professionals) helped her and other makers feel comfortable pursuing their ideas.

> Think[box] has great energy, and great people. It's a great place to test an idea, to explore an idea. You have the resources there, you have the materials there. You also have the knowledge there. The staff members are super great. They are very knowledgeable, very friendly. They also have working professionals there who are willing to help you with your project. They seem fun and are interested and invested in the project you are working on. I have met architects and engineers there.

Another student user that I interviewed was Ryan, who was in his junior year studying art at the Cleveland Institute of Arts (CIA). I have learned through Ryan that the CIA had a unique collaborative relationship with Case Western Reserve University, and that students from both institutions often collaborated, including working together in the Think[box]. Ryan was grateful for resources that he received as a CIA student through the makerspace.

> I have a one-year grant through the Think[box] so they supply me funding to support my own independent project. I will go there to 3D print, I will go there to laser cut. I will go there to just do general manufacturing. Sometimes I will just go up there to talk with other people and to see what they are working on. It's an interesting environment. It's just really fun to be up there.

When asked for his opinion on the kind of collaboration fostered through the Case Western-CIA collaborative initiative, Ryan noted that such effort is plausible because it brought artists and engineers together.

> Think[box] is very crucial for my academic development. At the Cleveland Institute of Arts, sometimes it is very dense there with artists, and you are not exposed to engineers, to makers. It is nice to get out of there. This is one of the reasons I chose to study at the CIA—it's because of its relations to Case Western and the Think[box]. There are about 500 students at the CIA, and I know there are a couple of foundation classes that push students to the Think[box]. Maybe 10% of CIA students make it over there before they graduate.

As with Nicola, Ryan was grateful for the resources made available through Think[box] and that students did not need to pay for most materials (e.g., plywood, filaments). He was also thankful for the expertise offered through the makerspace staff that has helped him with his projects.

Similar to my experience at the Invention Studio, I noticed a strong sense of community at Think[box]. The Think[box] website emphasized that the makerspace aimed to serve not just students at Case Western, but also the greater Cleveland community—business and non-business organizations alike. As it was evident through Ryan's experience, he benefited from an academic collaboration between CIA and Case Western. Nicola, too, pointed out that she was able to meet working engineers and architects in the makerspace. Think[box] has presented itself as a common space for communities beyond the university. The overall atmosphere of Think[box] established a sense of "entrepreneurship"—the way it was set up and run emphasized how individual projects could get ideated, designed, fabricated, and shipped as profitable products in a streamlined design process. This process followed the 7-step start-up route, which was also manifested in the makerspace physical structure (seven stories). With the additional incubator and project spaces, Think[box] stood out from the other two makerspaces I observed, articulating an entrepreneurial stance.

Comparing Three Makerspaces

Upon my return from Cleveland, the academic in me immediately resorted to comparing the three makerspaces, looking for similarities and differences among them. While I understand such exercise can be reductive and trivial, it has the potential to reveal a holistic view of well-functioning academic makerspaces and provide insights for those who are planning to deploy making as a pedagogical strategy in their courses. So, in this section I offer four key similarities and three differences across the makerspaces I visited and describe their implications for a maker-based pedagogy for technical communication.

First, the most identifiable similarity among the three makerspaces I observed was that they all served mainly engineering students and faculty. When finalizing my sites, I have worked to ensure that all of the makerspaces I study would be openly accessible so as to avoid disciplinary bias in how they are set up and operated. However, even though the publicity about the Anderson Labs makerspace appealed to its accessibility and service to the university as a whole, I later found out that it only served the CSE and engineering faculty. This seemed to be the case for the Invention Studio and Think[box] as well, although the two makerspaces did not limit access to just students or faculty from a particular college or department. I argued that they still primarily catered to engineering students and faculty because of the nature of the institutions themselves (both Georgia Tech and Case Western Reserve University were known to be engineering schools).

Second, since all three of the makerspaces were relatively big and known nationally as models for emerging makerspaces, they were well-equipped in terms of the tools and materials made available to makers. They all had similar fabrication and manufacturing tools, workbenches, and collaborative spaces where makers could meet and discuss ideas. The availability of the tools and the layout of the spaces were what made these makerspaces unique learning environments. The bias-toward-action learning philosophy that was exemplified by the respective makerspaces encouraged makers to put their ideas into tangible, testable forms early rather than getting stuck in the discussion of their ideas. These spaces were clearly designed with a design thinking philosophy, where failures were celebrated as part of the design process. The student interviewers all spoke to this notion when asked about their experience in materializing their design ideas.

Third, the design of the makerspaces fostered horizontal, or peer-to-peer learning. My student informants had all noted that they found values in working from their respective makerspaces in terms of learning from other makers in the space. Reciprocally, they all offered guidance or advice to their peer makers whenever they were asked for help. This kind of learning seemed desirable as students are typically less intimidated by their peers compared to their instructors. As peer mentors, makers could also become more proficient in a tool or a making process, helping them to better teach others.

Finally, I noticed that all three makerspaces had significantly active student involvement in its core operation. In these makerspaces, there were student groups or organizations that either helped run the facility or used it to perform learning activities that benefited the university at large. For instance, at UMN, student clubs like Tesla Works and Design U were student-led groups that hosted annual university-wide make-athons that took place in the Anderson Labs. At Georgia Tech, there was an official student club for the Invention Studio that organized a similar design competition. Georgia Tech students also served as board members and were trained to become Prototype Instructors or Prototype Masters who then volunteered in the makerspace. I was informed by student makers at the Invention Studio that all tools and technology purchases were requested by students and the affiliated faculty only signs off on the purchase requests. Lastly at Think[box], students were paid as workers and technicians in the makerspace. During my site visit, I found no professional technicians at Think[box]; they were all staffed by students. Overall, all three sites appeared to be extremely student-focused, even more so than traditional student learning facilities like university libraries or writing centers. This kind of student involvement (and investment) could be a great model for traditional learning spaces.

In terms of differences, I first noticed the three makerspaces were of different sizes, and they occupied their respective campuses in different ways. With more than 50,000 square feet, Think[box] was the largest among the three sites I visited, followed by the UMN Anderson Labs at 10,000 square feet, and the Invention Studio at 4,500 square feet. While the size of the makerspace does not represent its prominence or success, they do require different operational procedures and run on different budgets. I learned that the UMN Anderson Labs relied on a generous donation and were administered by CSE, one of the larger college units in the university system. Similarly, the Invention Studio was supported by a larger academic unit at Georgia Tech, the George W. Woodruff School of Mechanical Engineering. In contrast, Think[box] was an independent player. In its *Playbook*, Think[box] described the importance of engaging faculty, alumni, and key university players, as well as external partners to create an "ecosystem" that would support a standalone student-serving facility.

Another difference in these three makerspaces was what I would refer to as their persona. If I were to consider each of them as individuals, I felt as though I had made three different friends, each with unique personality and character. The first friend, the Anderson Labs, was very much focused on the manufacturing process, rather than conceptualization or entrepreneurship. I would refer to this friend plainly as "the shop." My second friend, the Invention Studio, came across as more developmental. I would call this friend "the design space." Makers in the Invention Studio were seen tinkering and prototyping using both digital fabrication as well as manufacturing tools. However, there was less welding and more 3D printing and electronic circuitry there compared to my "shop" friend. Lastly, my third friend, the Think[box], would be someone I refer to as "the entrepreneurial center." It was apparent in its presentation and publicity that this third friend focused on turning prototypes into start-up products. The entrepreneurship and incubator floors in the Think[box] building were physical manifestations of this ideal.

The last main difference among the three makerspaces was their level of community engagement. They had varying levels of engaging external entities such as business organizations and sponsors. The Invention Studio made it obvious that most student projects were sponsored by businesses around the area. Brand names and company logos could be seen on banners and posters that were hung around the makerspace. The UMN Anderson Labs, on the contrary, had no visible showing of corporate investment in its makerspace. For Think[box], community engagement meant not just bringing corporate sponsors to student projects, but also inviting them to use the makerspace for their own projects. The entrepreneurship and incubator floors in the Think[box] building were where businesses could rent temporary workspaces to create their own start-up initiatives.

Making as Design Thinking: Opportunities for Technical Communication

My visits to the three academic makerspaces had reinforced my confidence in the benefits of making for teaching and learning, and I argue they offer advantages for technical communication pedagogy as well. Although none of the student interviewees and makerspace staff members I spoke with had a technical communication background, they have all demonstrated how making can be a design-driven problem-solving strategy that could enrich learning, particularly when the subject matter is one of problem-based study—i.e., technical communication (see my contentions in Chapter 1). Throughout this book, I continue to demonstrate that making, design thinking, and problem solving are inherently intertwined, and technical communicators are increasingly finding themselves in situations that require such thinking and doing. From my observations at these makerspaces,

I noted that students were not only creating innovative solutions to solve problems, but also generating communicative artifacts such as process memos or journals, product documentations, and user guides or instructions to accompany their inventions. While these artifacts were almost always overshadowed by the actual inventions themselves, they are a crucial part of the innovation and problem-solving process that required technical communication competencies to deliver a total—what some designers call, *full-stack*—solution.

Making has the potential to foster meaningful learning through collaboration and peer mentorship. All of the students I interviewed at the three makerspaces revealed that they relied on other makers in the makerspace when building their projects. Although some of their projects were independent, these students revealed that they all have asked other makers for help at some point during their work in the makerspace. Whether they were needing help with a specific technology or simply asking for another person's perspective, they noted how peer feedback was helpful for the development of their work. One student has especially noted that by exposing her work to other makers to the makerspace, she "let other makers critique her work" and gained critical perspectives she wouldn't usually receive in a classroom setting.

In a similar way, making can also teach our students to be mentors for their peers. Most of the students I interviewed noted how working in their makerspace has taught them to be learners who are motivated to succeed in their respective projects *as well as* helping others in the same space. They expressed a sense of achievement and when they were able to assist those in need. Ryan, for instance, noted that his involvement at the Think[box] has taught him to be sensitive to other makers in the makerspace and be helpful whenever possible. At the Invention Studio, Nicola noticed that she learned from helping other student makers in the space because she did not believe in just one correct way in doing anything. From assisting others, she learned from other makers' mistakes and was able to apply those experiences in her own project. In some cases, this peer level learning has resulted in bigger, continued collaboration such as Ryan's project that bridged the CIA and Case Western.

As spaces for productive experimentation, or tinkering, makerspaces offer students a unique environment to acquire new skills and experience. And because making promotes a mindset that celebrates failure as part of problem solving, students may be encouraged to try new solutions or approaches. For technical communication pedagogy, this can foster the kind of entrepreneurial thinking that Bay et al. (2018) has forwarded to our field. Makerspaces and their associated training programs, such as the Prototype Instructor and Prototype Master certifications at the Invention Studio, can provide students with structured introduction to technologies that can take their projects to the next level. This ongoing professionalization spirit bolstered by making can strengthen students' skill sets, especially for the world of technical communication that is constantly influenced by changing technologies.

This project has allowed me to gain firsthand knowledge and experience in three different makerspace settings. These insights have proven to be critical to the latter part of my doctoral project, where I designed and deployed a maker-based technical communication course. I discuss the deployment of this course and its results in Chapter 4, following the intermediating arguments for social innovation in Chapter 3.

Summary and Takeaways

In this chapter, I have synthesized current discussions that indicate the need for technical communication pedagogy to turn to design-centric and material thinking. To demonstrate an opportunity for updating our pedagogy, I have provided a historical account of the fast-growing Maker Movement. Through a survey of the roots and technology of making, I contended that maker practices could reinvigorate technical communication pedagogy. By reporting a study of three academic makerspaces in the United States, I showed the impact of maker culture on higher learning via the cases of engineering and design. Using the findings from student interviews and my own in-situ observations, I reiterate the intertwined relationships between making and design thinking and problem solving in technical communication. Key takeaways from this chapter are:

- Making as an experiential, problem-based learning component can be a strategic direction for technical communication pedagogy.
- The Maker Movement can be traced back to the industrial revolution and the DIY culture propelled by self-sufficiency and adhocism.
- Academic makerspaces are sites for fostering collaboration, peer mentorship, and community engagement.
- Making can serve as a springboard for technical communication students to practice design thinking and learn new technologies.

Learning Activity: Transforming a Classroom Into a Makerspace

A lot of my colleagues from college campuses across the country do not have access to a local makerspace. They often ask me how it would be possible to simulate a maker culture without the physical makerspace. I see that as a design challenge that can be a meaningful exercise for instructors and students alike. By transforming your current learning spaces into a makerspace, you may learn to see spaces differently and pay attention to resources.

The first step in transforming your classroom into a makerspace is to consider the existing assets in your space. Can your classroom furniture and technologies—desks, chairs, projects, screens, computers, speakers, etc.—be moved or rearranged? Where are the power sources? Sometimes, all it takes to reconfigure a

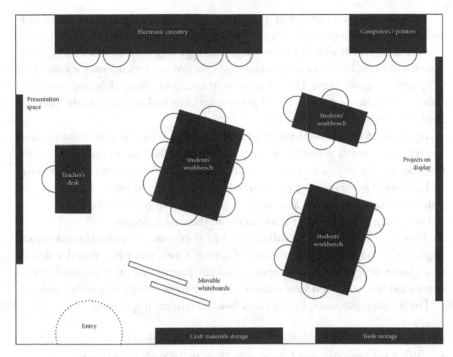

FIGURE 2.1   A simple makerspace setup in a classroom

classroom is a few power extensions. Figure 2.1 shows an example of setup that can be conducive to making.

Most instructors and students do not have the funds to upgrade their classrooms so the ideal transformation would be done with low to no budget. One strategy I have learned from some creative minds is to tap into the university waste management department. At both my former and current institutions, there is a waste site where our engineering departments and other physical plant units of the university would deposit unwanted materials. More often than not, you may find some old desks, benches, or scrap woods that can be repurposed.

Assemble a team of builders (motivated students!) and set them on a mission to build simple workstations like a workbench, crates for storage and organization, etc. This is a good practice in user-centered design since you are having the users contribute to the creation of an artifact they would end up using themselves later. Pro tip: put wheels on your furniture so you can easily move them into different setups as necessary (or return the classroom to its original setup for the next teacher).

The next step is to put out a call for donations. Create a flyer. Send out an email. Post a tweet. Anything from paper towel tubes, yarns, buttons, fabrics, LEGOs (a lot of them) to circuitry components like conductors, LED lights, and sensors are welcomed. Hand tools like hammers, hot glue guns, and screwdrivers can be easily found in a rummage sale nearby.

Now if you have a little bit of money, consider investing in small power tools as well as physical computing electronics like Arduino (www.arduino.cc/en/guide/introduction) and Raspberry Pi (www.raspberrypi.org/). These can be bought by your institution's library or tech center, and checked out by students for your class. 3D printers are overrated. Do not waste your time and energy debating with your college administrators if you could get one for your makerspace because they are hard to maintain and slow to run.

In terms of design software, a subscription to Adobe Creative Suite is ideal. However, Microsoft PowerPoint—and students usually have access to it—can do basic layout design just fine. Hopscotch (www.gethopscotch.com/) is a popular and accessible programming application for students. If you have a 3D printer (no judgement here), you may also want to get Tinkercad (www.tinkercad.com/), a 3D modeling software, to let students create their own designs.

There you have your basic makerspace in a classroom. A sustainable makerspace requires a sustainable funding model, of course. Once you have created a decent space, invite your college administrators, students' parents, and other potential collaborators and funders to visit your creation, and urge them to support your makerspace.

For learning purposes, I offer the following reflection questions:

- Who are your main audience/users of this makerspace?
- What resources do you have connections to, or wish to tap into?
- What can a makerspace do for your course/pedagogy as well as other courses?
- How might your transformation inspire actions by your students and colleagues?
- Who should assume ownership over the makerspace?

Note

1. There has been critique over the use of the term *maker* as an emasculating term that privileges white, male, able-bodied creators versus their counterparts. Art professor Diane Willow at the University of Minnesota has shared with me a story wherein a female colleague who invented the technology for the LilyPad e-textile circuit board was only ever noted as a "crafter" in mainstream reports rather than a "maker" due to her identification as female. For the purpose of clarity and consistency, I use "maker" as an androgynous term to represent all creators who embody the spirit of the Maker Movement.

References

Australian Institute for Teaching and School Leadership. (2014). Learning through doing: An introduction to design thinking. Retrieved from www.aitsl.edu.au/docs/default-source/default-document-library/aitsl-learning-through-doing-introduction-to-design-thinking.pdf?sfvrsn=e0bdec3c_0

Barrett, T., Pizzico, M., Levy, B., & Nagel, R. (2015). A review of university maker spaces. In *Proceedings of 122nd ASEE annual conference and exposition* (pp. 1–17). Washington, DC: American Society for Engineering Education.

Bay, J., Johnson-Sheehan, R., & Cook, D. (2018). Design thinking via experiential learning: Thinking like an entrepreneur in technical communication courses. *Programmatic Perspectives, 10*(1), 172–200.

Breaux, C. (2017). Why making? *Computers and Composition, 44*, 27–35.

Brown, J., & Rivers, N. (2013). Composing the carpenter's workshop. *O-Zone: A Journal of Object-Oriented Studies, 1*(1), 27–36.

Cargile Cook, K. (2002). Layered literacies: A theoretical frame for technical communication pedagogy. *Technical Communication Quarterly, 11*(1), 5–29.

Cavalcanti, G. (2013). Is it a hackerspace, makerspace, techshop, or fablab? *Make:*. Retrieved from https://makezine.com/2013/05/22/the-difference-between-hackerspaces-makerspaces-techshops-and-fablabs/

Craig, J. (2014). Makers and makerspaces: Teaching composition in a creative economy. Retrieved from http://unwrite.org/pearson/

Cumming, E., & Kaplan, W. (1991). *The arts and crafts movement.* New York, NY: Thames and Hudson.

Dougherty, D. (2012). The maker movement. *Innovations: Technology, Governance, Globalization, 7*(3), 11–18.

Educause. (2013). Seven things you need to know about makerspaces. Retrieved from https://net.educause.edu/ir/library/pdf/ELI7095.pdf

Elam-Handloff, J. (2016). Making across the curriculum: DIY culture, makerspaces, and new modes of composition. Gayle Morris Sweetland Digital Rhetoric Collaborative (DRC). Retrieved from www.digitalrhetoriccollaborative.org/2016/03/03/making-across-the-curriculum-diy-culture-makerspaces-and-new-modes-of-composition/

Haas, C. (1996). *Writing technology: Studies on the materiality of literacy.* Mahwah, NJ: Lawrence Erlbaum Associates.

Hatch, M. (2013). *The maker movement manifesto: Rules for innovation in the new world of crafters, hackers, and tinkerers.* New York, NY: McGraw-Hill Education.

Jencks, C., & Silver, N. (2013). *Adhocism: The case for improvisation.* Expanded and updated edition. Cambridge, MA: Massachusetts Institute of Technology Press.

Make: magazine. (n.d.). Maker media. Retrieved from http://makezine.com

Maker Faire: A bit of history. (n.d.). Retrieved from https://makerfaire.com/makerfairehistory/

Maxigas, P. (2012). Hacklabs and hackerspace-tracing two genealogies. *Journal of Peer Production, 2*, 1–10.

Morozov, E. (2014, January 13). Making it: Pick up a spot welder and join the revolution. *The New Yorker.* Retrieved from www.newyorker.com/magazine/2014/01/13/making-it-2

Prior, P., & Shipka, J. (2003). Chronotopic laminations: Tracing the contours of literate activity. In C. Bazerman & D. Russell (Eds.), *Writing selves, writing societies* (pp. 180–238). Fort Collins, CO: The WAC Clearinghouse. Retrieved from http://wac.colostate.edu/books/selves_societies/prior/

Rose, M. (2014). The maker movement: Tinkering with the idea that college is for everyone. *Truthdig.* Retrieved from www.truthdig.com/articles/the-maker-movement-tinkering-with-the-idea-that-college-is-for-everyone/

Roy, D. (2016). Task-based technical communication with 3D-printing-based initiatives in a foreign language teaching context. Paper presented at the Annual Technical Communication Symposium, Kyoto, Japan.

Sayers, J. (Ed.). (2017). *Making things and drawing boundaries: Experiments in the digital humanities.* Minneapolis, MN: University of Minnesota Press.

Sheridan, D. (2010). Fabricating consent: Three-dimensional objects as rhetorical compositions. *Computers and Composition, 27*(4), 249–265.

Shipka, J. (2011). *Toward a composition made whole*. Pittsburgh, PA: University of Pittsburgh Press.

Stockman, A. (2016). *Make writing: 5 Teaching strategies that turn writer's workshop into a maker space*. Cleveland, OH: Times 10 Publications.

Tham, J. (2020). Engaging design thinking and making in technical and professional communication pedagogy. *Technical Communication Quarterly*. Published online first. Retrieved from www.tandfonline.com/doi/full/10.1080/10572252.2020.1804619

Trimbur, J. (2000). Composition and the circulation of writing. *College Composition and Communication, 52*(2), 188–219.

What is a makerspace? (2015). Retrieved from http://pages.vassar.edu/makerspacetalk/2015/10/21/what-is-a-makerspace-2/

Yancey, K. B. (2004). Looking for sources of coherence in a fragmented world: Notes toward a new assessment design. *Computers and Composition, 21*(1), 89–102.

3

SOCIAL INNOVATION

Designing Humane Technical Communication

Overview: Focusing on the important role technical communicators play in user research, this chapter highlights the connections between design thinking and social advocacy, which is now infamously framed as the interdisciplinary effort of social innovation. This chapter ties this effort with the work of technical communication and contends that technical communicators should be leaders in human-centered problem solving. Through case examples and interviews with industry experts, this chapter presents scholarly as well as industry perspectives on social innovation by technical communicators. These examples demonstrate the immense opportunities for technical communication, as a discipline, to assume leadership in social innovation initiatives and to do so by bridging classrooms and industries.

Technical Communication as Changemaking

Among the key lessons I learned as a copywriter in my pregraduate school years is the role of strategic communication in changemaking. Under the mentorship of an ever-so-patient account director and a fiercely artistic creative director (Terence and Wong!) at a boutique ad agency, it was instilled in me that the purpose of advertising goes beyond calling attention to a product or service; it is a catalyst for culture and change. We serve our clientele by helping them make smart yet ethical decisions, earn trust, and create positive public engagement. As a writer, my work included collaborating with the creative team—graphic designers, photographers, videographers, web developers, and editors—to cultivate *actions* through meaningful messages. When executed strategically by clear, clever, and compelling rhetoric, I have witnessed the way good ideas bring on great results.

Shifting into the world of technical communication, I find many parallels between technical and marketing writing, in particular the common goal of changemaking. In their award-winning book, *Key Theoretical Frameworks: Teaching Technical Communication in the Twenty-First Century*, editors Angela M. Haas and Michelle F. Eble (2008) posited that modern technical communication pedagogies and practices require "social justice frameworks" that "explicitly seek to redistribute and reassemble—or otherwise redress—power imbalances that systematically and systemically disenfranchise some stakeholders while privileging others" (p. 4). Haas and Eble, along with the contributors to the collection, offered a myriad of perspectives into the ways technical communication practitioners and teachers can affect social change by paying attention to the cultural imperatives of their work and pedagogy today. Specifically, Haas and Eble forwarded a set of foundational approaches to social justice through technical communication that's worth quoting in full:

- All technical communication has the potential to be global technical communication. Even if one works in/for a local organization, the technical communication of those outside the organization could shape the technical communication that transpires within, not to mention that stakeholders and/or users of that technical communication may come from diverse global locations.
- Social justice is both a local and global necessity. This means that contrary to rhetorics of national exceptionalism, the United States, "first-world," and Western countries could also benefit from social justice approaches to technical communication.
- International and intercultural communication happens outside of non-Western and non-US contexts (and without Western and "first-world" interlocutors). Moreover, these cases, their stakeholders, their technical communication—thus, cultural and rhetorical—work, and the power dynamics therein are worthy of our study.
- International technical communication happens within the United States. There are over five hundred sovereign indigenous nations independent from the United States but are located within United States national borders. And this international technical communication can and does happen independent from United States and other "first-world" involvement.
- International and domestic technical communication is all a matter of rhetorical perspective. A case study of Chinese technical communication, for example, is not international technical communication for Chinese technical communicators.
- Intercultural technical communication happens within and across national borders given ethnic and other cultural diversity.
- Although social justice begins at home, it's important to understand the relationships between local and global injustices. Certainly, we should consider our agency as technical communicators in light of the social injustices within our own communities rather than positioning ourselves as rhetorical missionaries for Others. But we should also study the patterns and trends across

and between local and global stories of injustice so that we may better identify, analyze, and redress the ideologies, institutions, stakeholders, and rhetorics that sponsor them—and to more effectively form intercultural technical
communication teams to do so.

- Social justice includes justice for the environment, as injustices against any
  living species (not just humans and non-human animals) should impact the
  social. Moreover, many non-Western epistemologies understand non-human
  actors as social beings.
- Social justice benefits everyone. Working to achieve or restore equity for
  one population or community does not require anyone with access to those
  rights to relinquish them—quite the opposite actually. For technical communication, specifically, equity means fair and just access to and representation
  in scientific and technical communication for *all* stakeholders.

(Haas & Eble, 2018, pp. 10–11)

Taking heed from Haas and Eble's disposition, Rebecca Walton, Kristin Moore,
and Natasha Jones (2019) in their equally influential book—*Technical Communication after the Social Justice Turn*—positioned technical communicators as changemakers who should consider the core of their work as withstanding unjust issues
and resisting oppressive practices. Similar to my experience in advertising, good
communication—whether technical or creative—should promote actions toward
a good cause. But in order to arrive at such an ideal, technical communicators
need a combination of mindset and skillset that allows them to address the problems they encounter in the ongoing process of cultivating change.

Noted in the previous chapters, I subscribe to a problem-solving characterization for technical communication as popularized by Johndan Johnson-Eilola
and Stuart Selber (2013). Technical communicators in each of their respective
industries—medical, legal, computer engineering, etc.—are solvers of *sociotechnical* problems. Technical tasks are staple work in the profession. Many novice
professionals entering this industry expect to work on different aspects of content
creation and delivery. Rather, the social part of technical communication is arguably more intricate than its technical counterpart. Social problems are complex,
political, culturally situated, and incomplete; they are what Rittel and Webber
(1973) called "wicked problems," or problems without objective definitions nor
optimal solutions. Due to the perplexing conditions of our global society and
the increasingly convoluted nature of technical communication practices (intertwined with adjacent professions) in the face of evolving technologies, technical communicators are challenged with unprecedented problems that require
not-yet-available solutions. In other words, technical communicators must innovate solutions. To do so, we need to "look outside the tech comm bubble," as
Amazon technical writer and blogger extraordinaire Tom Johnson[1] (2015) put it.
We should consult with other fields, even those that may seem too remote from
our immediate interests or practices. Alas, as reality shows us, technical communication work should always-already be interdisciplinary.

During the development of this book, our world was struck with two major crises—a worldwide health pandemic brought about by COVID-19 and a global uprising stirred by systemic racism. The public role of technical communicators has since become more prominent than ever. We are called to take on innovative approaches to affect change. And this changemaking required empathy, creativity, and strategy. For example, during the pandemic, technical communicators across the world have created different user-friendly data visualizations and infographics to educate citizens about the development of the deadly virus and steps to ensure personal hygiene and safety. Posters, memes, videos, and other forms of media have been deployed to educate the masses. Strategic communication professor Curtis Newbold (2020), the creative mind behind The Visual Communication Guy website, has produced freely downloadable graphic flowcharts for restaurant owners, schools, and other workplace programs to guide their operations during a time of anxiety. Other technical communicators have occupied social media spaces to perform similar educational efforts through creative and rhetorical methods.

Meanwhile, as the health pandemic called for technical expertise, the strife for racial justice demanded attention by technical communicators to the social dimensions of their crafts. Technical communicators—including UX researchers, content strategists, instructional designers, etc.—can participate meaningfully in social advocacy through their professional practices. Salesforce senior design researcher Vivianne Castillo has openly called UX practitioners to speak out about social problems such as "privilege, racism, homophobia, white supremacy, xenophobia, etc." (quoted from tweet in Figure 3.1) because UX relies on

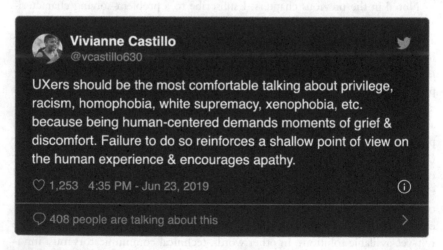

FIGURE 3.1 A tweet by Vivianne Castillo on June 23, 2019 asking UX professionals to talk about privilege, racism, homophobia, white supremacy, xenophobia, etc.

Source: Used with permission

human-centered values. These values should be translated into actions through technical communication activities and advocacy. Indeed, technical communicators need to do more in paying attention to users as marginalized people. There are tools to do this work, and this chapter aims to demonstrate those tools from the design thinking standpoint.

A Call to Advocacy

In one of his many public lectures to engineering students and designers, former Apple, Inc. VP Don Norman (2014) called out the insensitivity in computer systems developers and designers today when it comes to user advocacy. "Efficiency and functionality aren't the same," proclaimed the former professor of cognitive psychology and computer science. Norman reiterated a point he had made since the 1980s that systems designers (including interface designers) need to advocate for user needs. Designers must consider user goals, understand user expectations and behaviors, and most importantly, involve the user in the design and development process. User-centered design is born of this heuristic. In "Ethics of Engagement," Michael Salvo (2009) highlighted the distinction between user observation and user engagement. Through his case studies, Salvo observed that user engagement invokes user-centered design through dialogic interactions and participatory design methods. When users claim ownership and contribute directly to the actual design process, the product would result in more desirable user experience. In addition, the U.S. federal resource page Usability.gov (n.d.) stated that benefits of user-centered design also include reduced user errors, reduced developmental cost, and increased sales.

However, since technology and its design are never neutral nor objective, technical communicators have argued that user advocacy should go beyond the considerations for efficiency and satisfaction. Evidently, our field has always taken on issues of ethics and cultural ideologies to combat inequitable practices. Steven Katz's (1992) landmark critique of the "ethic of expediency" in technical documentations has shed light on the problems with deliberative rhetoric—typically taught in college writing courses as strategies for persuasion—disguised in style, logic, and affective appeals. The obsession over proper formatting, logical presentation of information, and technical accuracy in technical communication can create a blind spot to the social and political implications of technology. Sam Dragga and Dan Voss's (2001) study of data display and other technical illustrations showed a similar concern. Technical information tends to mask humanistic values and strip away compassion for the subject matter. To rectify, technical communicators should "adapt the technical to the human" (Dragga & Voss, 2001, p. 272). I agree and challenge technical communicators to take it upon themselves the responsibility of standing by users and being their strongest advocates.

User advocacy must include paying attention to minoritized and marginalized user communities, inviting their perspectives and contributions, and empowering their involvement in the building of technologies. A telling example can be found in the case of racist algorithms. Joy Buolamwini, while a grad student at MIT, had researched and fought against racial bias in machine learning. Noticing that facial analysis applications would not detect darker skin tones and facial structures, Buolamwini (2016) urged designers to recognize coded bias and how users can help mitigate such problematic design. Similarly, Safiya Noble (2018) in her best-selling book, *Algorithm of Oppression*, revealed the influences of human biases and values on technology design. Following Buolamwini and Noble's work, Cruz Medina and Octavio Pimentel (2018) have curated in their open-access digital book, *Racial Shorthand*, a series of culturally relevant examples that demonstrate the erasing of non-dominant cultures through oppressive technology design. These examples showed negligence on the part of designers and communicators, and reinforced the importance of centering justice work in technical communication.

An oft-cited set of ethical guidelines,[2] the Society for Technical Communication (STC) code of ethics states:

> We seek to promote the public good in our activities. To the best of our ability, we provide truthful and accurate communications. We also dedicate ourselves to conciseness, clarity, coherence, and creativity, striving to meet the needs of those who use our products and services. We alert our clients and employers when we believe that material is ambiguous.
>
> *(Ethical principles. Society for Technical Communication, 1998)*

The "public good" in this case should highlight social justice components among other principles like professionalism, honesty, and quality that make up the fabric of technical communication ethics. We need to consider our craft and invention in terms of their social impact on different stakeholders and the larger community. We must be sensitive toward the effects of our work especially to those who lack the power or autonomy to make choices. We must empathize with the real-world conditions our users experience and expand the promotion of the "public good" to include combating injustice through technical communication activities. There are exemplary cases that can teach technical communicators and user experience designers some lessons on user advocacy; the next section features three instances of user-centered social innovation that are informed by design thinking.

Where the Rubber Hits the Road: Social Innovation Examples

Stanford's Center for Social Innovation faculty members Sarah Soule, Neil Malhotra, and Bernadette Clavier (n.d.) defined social innovation as "the process of developing and deploying effective solutions to challenging and often *systemic social and environmental issues* in support of social progress" (n.p., emphasis added). The genesis of any social innovation project is often an exigence brought about by social injustice. In "Disrupting the Past to Disrupt the Future," Natasha Jones, Kristen Moore, and Rebecca Walton (2016) effectively traced the social justice movement in technical communication to the field's early influences by the social paradigm and its growing research trajectory toward inclusivity. Reasonably, this trajectory lands on social innovation as one of the (ante)narratives that situate technical communication as social activists and user advocates. To showcase the role of design thinking and social innovation research in our field, I spotlight three case examples here as a demonstration of advocacy through socially responsive innovation.

Example 1: Project EMAR—Participatory Design

Human-computer interaction has been a growing area of study in technical communication since the proliferation of computer-driven communication and information delivery. The rise of physical computing, boosted by the Maker Movement, has caused increased attention to robotics from our field. For good reasons, many engineering and human-robot interaction projects have focused on designing robots for the elderly. However, there is a lack of attention to a significant population underserved by robotics research—children and adolescents. Technical communication and user experience researchers Emma Rose and Elin Björling, in collaboration with robotics scientist Maya Cakmak, have led a team of student researchers at the University of Washington to design a social robot that captured adolescent stress levels. Funded by the National Science Foundation, the Ecological Momentary Assessment Robot (EMAR) project is an exemplary application of design thinking. Specifically, the EMAR project demonstrated the necessity of participatory design as informed by design thinking principles. The EMAR team engaged local high schools and sponsored activities like a robot design challenge to let students envision and create a robot interface with which they prefer to interact ("Project EMAR," n.d.). Teens were given foundational knowledge in human-centered design and user-interface design, and were invited to engage in the ideation and prototyping processes through collaboration (team-based competition).

The key to successful participatory design is meaningful engagement, as Björling and Rose (2019) reported. As co-designers, participants should feel that their contributions would have a meaningful impact on the outcomes of the project. More importantly, participatory design maintains "an ethical and meaningful stance with the population they intend to serve" (Björling & Rose, 2019, p. 2). It motivates researchers and designers to focus on the social implications of their project rather than mere productivity or profits. Project EMAR also embodied the spirit of social innovation as it situated social problems at the center of design, while bridging user experience design as a technical communication practice with robotics design as an area of interest for emerging technologies researchers (Rose & Björling, 2017).

Example 2: inControl—Rapid Prototyping

Social innovation uses design as a platform to call out oppressive cultures and practices. In technical communication and across adjacent fields, ableism is often overlooked and manifests in many of our products and processes. Fortunately, the critical tradition in our discipline has constantly reminded us to reflect on our privileges and advocate for more inclusive practices. Accessibility and universal design scholars have been the forerunners in social justice advocacy work in technical communication. Writing teacher Deborah Wood and industrial engineering instructor Janice Mejia have co-advised a small team of students in the design thinking and communication program at Northwestern University to address accessibility issues in game design. Within merely 10 weeks, this student-led project has built a prototype for a single-hand operated video game controller, inControl, that showcased the power of design thinking for social innovation ("inControl: A video game controller," n.d.).

The inControl team redesigned the joystick and buttons on the game controller to meet the needs of patients with hemiplegia. The project relied on rapid-prototyping tools including 3D modeling software, laser cutters, 3D printers, and electrical circuit boards to build a testable prototype. Given the short time it took from problem definition to prototype testing, this project can be used as an exemplar for technical communicators who work at the intersection of accessible design and game studies. It demonstrated the practicality of design thinking for addressing a social issue about which our field has been concerned. It also highlighted the importance of prototyping as a process to materialize ideas to achieve actual social impact. Per the project page for inControl, the student designers, under faculty assistance, have filed a provisional patent application for this controller design. From a pedagogical standpoint, design thinking adds an entrepreneurial layer to the conventional learning experience of students, giving them a special edge into their future career.

Example 3: Google Glass—Why We Must Begin With Empathy

Unlike the previous examples, the third instance featured here is a failed case of social innovation. I offer here a critique of a short-lived technological breakthrough in wearable computing, Google Glass, which had garnered global attention between 2013–2015. Many technical communicators and user experience designers were particularly intrigued by the introduction of the unprecedented wear-on-your-face computer, when Google's very own co-founder Sergey Brin took stage at a major TED Talk event in 2013 and gave his audience an imaginative ride on the roller coaster of computing's future. Scholars and practitioners alike wanted to know how this new device might influence or change the way users engage with screens and non-screen interactions (audio, haptic, etc.).

Arguably the first major consumer wearable device on the market, Google Glass offered users the ability to look up information online, take images and videos, translate foreign languages, and even live-stream events while keeping the wearer's hands free from a bulky device. What would that mean for technical instructions? What about virtual communication and collaboration? For many industries, Google Glass promised a leap into a future where action-oriented leisure meets the possibility of sensation without mediation (Pfister, 2019).

While I was a member of the University of Minnesota Wearables Research Collaboratory, I had examined the potential use of Google Glass in digital literacy development. As part of the collaboratory's collective research, we noted the affordances of the device but also recognized its lack of focus on users' *actual* needs. Google Glass offered the aforementioned functionality with a popular narrative that centered around revitalizing human's social nature by freeing them from their phones. In his TED Talk, Brin asked: "In addition to potentially socially isolating yourself when you're out and about looking at your phone, it's kind of, is this what you're meant to do with your body?" Yet, as many smartphone users might concur, there is more to why we use phones the way we do—like how we kept our screens to ourselves, how we wanted to know if others were recording us, how we used our phone as an escape mechanism when we wanted to avoid personal interactions. . . all of which were social implications of technology that the Google Glass project failed to empathize with. The design of Google Glass prioritized efficiency, but it disregarded the contexts in which users might perform the different actions it enabled. All in all, the Google Glass example serves as a cautionary tale wherein social innovation designers may learn about the importance of empathy. Design thinking begins with empathy because it is the foremost crucial element in the design process. As the primitive Google Glass case showed us, innovation should start with empathy toward users or it would not succeed; functionality and features of technology must address existing problems as identified by users rather than the imaginations of designers.

Implications of Social Innovation for Technical Communication

From the previous examples, we may observe the influence of social innovation on the work of technical communication. As evident in the inControl and Project EMAR instances, social innovation differs from traditional business models that stem from major research and development centers or technocracies of central administrations (private or public). For technical communicators, social innovation expresses the vitality of a pluralist approach to invention that represents hopes and aspirations that challenge oppressive or exclusive concepts of innovation. Design thinking actualizes these motivations by providing tools to democratize the innovation process, through methods like participatory design and rapid prototyping.

Social innovation helps keep technical communication as a user- and human-centered profession. As the consistent narrative in modern technical communication theories and manifestos urge practitioners to be user advocates, I have argued that user advocacy extends beyond generating user stories and personas; user advocacy in our world of injustice and oppression means recognizing equal power dynamics, actively combating unjust treatments through design and communication, and positioning justice at the center of our practice. Powered by the first step of design thinking—empathy—social innovation can achieve true user advocacy by keeping the starting point in design on users. Indeed, good technical communication already does that, a social innovation perspective maintains this important mindset and helps practitioners to prioritize the social over the technical.

As the Google Glass example revealed, failing to work with social contracts can defeat even the most groundbreaking technological invention. A social innovation mindset emphasizes a view of technological progress that is led by the human dimension of invention. As user advocates, technical communicators serve as critical translators of user needs—their cultures, backgrounds, living conditions, and everyday struggles—for design teams. Designers and developers should rely on these important human conditions in producing new technologies that would impact the lives of their users.

Technical communication as a matured and evolving profession can provide expertise in sustaining local social innovations. Current literature frames the reality of local social innovations as small and temporary, although they represent high value for their local contexts. The collaboration between technical communicators, robotics scientists, and educational communities presents an exemplary case for sustainable social innovation. The partnership brought together expertise as well as problem-based learning afforded by design thinking practices. The growing network of schools and students participating in the innovative initiative fostered the conditions for creativity and productivity. Educators and students participating in the design challenge activities earn design experience and transferable skills. Researchers facilitating and observing the process gain insights to user needs. The collective outcomes amount to the iterative design of the innovation—a social robot—and its results could be scaled to different requirements with implications for the direction of future funding.

The role of technical communicators in these social innovation cases may differ, but they revolve around entrepreneurial readiness. No longer individuals who write or manage content, nor a siloed department for translating complex documents, technical communicators are an important part of the changemaking process where they actively share leadership in social activism and user advocacy. To survey the current professional climate of social innovation in the workplace, I have taken on a study of technical communication practitioners. I sought to understand how different technical communicators saw their roles in the big picture of social advocacy and innovation, and the influence of design thinking in their line of work.

Views From the Industry

Similar to the approach I took to understand the effects of making and maker-spaces on students in the previous chapter, I gathered insights from five industry practitioners through semi-structured interviews as a way to identify the practicality of social innovation in the commercial world. Interview questions included:

- Please briefly describe your professional background and current role at work.
- How would you describe the role of design—as you understand it—in your scope of work?
- What connections do you see between technology, technical communication, design, and innovation?
- How would you describe the role of technical communicators in relation to your line of work?
- What do you think is the role of technical communicators in affecting social innovation and positive social change?

The practitioners, each with different backgrounds in technical communication, shared a common trait that is a vision toward the growth of social innovation in their respective practices. In what follows I summarize the key implications of social innovation as seen by these practitioners, using pseudonyms to protect the informants' identity.

Social Innovation in IT

I relied on my existing professional network to locate my interviewees. Through my alumni connections I managed to gather the attention of John, a graduate from my alma mater, who at the time of this study ran an independent leadership service business for IT professionals. I have learned about John's business in a 2016 conference for technical communicators where he gave a presentation on the use of data visualization in user research. To create a baseline for this study, I began all my interviews by asking the practitioners to describe their understanding of design thinking and its relation to their respective scopes of work. John considered design thinking to be a relatively new buzzword in the IT world although the focus on design as a social approach to solutions building has been the core of effective IT businesses.

"Design is about the overall infrastructure. It goes beyond the user interface into the entire chain of an application," John said, emphasizing the notion that design promotes a wholesome thinking, rather than just focusing on aesthetics or appearances. John also highlighted the importance of usability testing and user experience studies in improving the design of any product. He identified these research skills to be the main contributions of technical communicators to their companies.

John noticed that content creation is only a small part of a technical communicator's job today; a majority of their work involves crafting and improving the design of the content—again, not just the visual presentation but the placement, customization, and fitting of the content in the overall scheme of the solution.

For John, technical communicators are well suited to wear the hat of a social innovator in their workplace. As liaisons to IT, engineering, and design teams, John believed that technical communicators occupy an important role that can facilitate real change. "Technical communicators have the ability to synthesize needs and feedback," John stated, and that ability is crucial in making sure products are user-centered and socially responsible.

Social Innovation in Medical Design

Outside of IT and computer technology development, one of the fields that have shown growth in hiring technical communicators is medicine. This can be observed through the sprouting of medical device agencies and the resulting surge in the need for UX designers and researchers. This industry is also flourishing thanks to the advancement in prototyping technologies, which afford lower design and development costs than conventional manufacturing processes. My second interviewee, Martin, was a technical communicator in medical software design. From him I learned about what it takes to actualize social innovation from an organizational standpoint.

Martin revealed that his company was a late adopter in the medical field in terms of user-centered design. Their prior design approach gave emphasis to usability testing but the designers later realized they needed to involve users from the start of the design process. To do so, Martin's company needed to revamp their design team structures. Using Agile project management methodology with a commitment to UX design principles, Martin's company revised the "waterfall" design practice into Agile teams. Unlike the sequential design process in the waterfall model, each team now consists of four to seven software designers and UX researchers who work together in each phase of the design process. "This allows teams to design their products iteratively," Martin said.

Martin saw empathy as the core of social innovation. Since their primary products were made for doctors, Martin said design thinking kept teams focused on a "by physicians, for physicians" mindset when designing products. Although not everyone on the team has a medical background, this mindset helped cultivate empathy in the design teams and enforced the practice of user engagement in every step of the design process. Since social innovation is about focusing on the living conditions of the user, Martin expressed, there has been an "accessibility push" in his company with a formal committee that assess all product designs and that was responsible for heightening awareness in emerging accessibility issues. Because design thinking is grounded in user-centered design, Martin believed that "design thinking certainly has staying power."

Social Innovation in Technical Documentation

Having spoken with practitioners in fields that were not traditionally considered as the primary career outlook for technical communication, I encountered Kristin, who identified herself a "through-and-through" technical writer at a multinational company that produces instrumentation software. Kristin worked in a writers' team that was responsible for creating documentations for multiple audiences, including developers, manufacturers, and end users. For Kristin, design thinking is equivalent to UX design. "Writers are user-centered designers!" Kristin proclaimed. While she considered the main job of a technical documentation writer to be generating and modifying textual content, Kristin saw design to be an integral part of that process.

Ultimately, the documentations that technical writers create would ship with the product and populate for users when they initiate the product, Kristin explained. The key to a great user experience is for writers to foster a designer mindset, where they concern themselves with user needs. "Design is to reduce frustrations," she put it simply. However, due to the size of her company and highly compartmentalized teams, Kristin revealed that products are tested by separate research teams and she did not always have access to the findings. This created challenges for writers who wanted to know the trends and behaviors in users. "Some products would have more focus on user experience design, but sometimes designers work behind closed doors," thus making it difficult for writers to understand the contexts of use, user needs, and customer experience of the product, Kristin said.

When asked about her view on social innovation, Kristin said that initiatives need to come from higher administration in a large company. Even though modern technical communicators may want to engage social issues and be stronger user advocates through their work, Kristin said writers may not always have a say in design directions if the company culture does not enable open sharing of user information. For Kristin, social innovation should be "an organizational priority," and not just at the individual level. For now, she reported that technical writers on her team mainly focus on complying with legal requirements in their documentation, rather than ethical issues that needed attention as well.

Social Innovation in Academic UX Services

My final informants were two UX experts working in an academic setting. In addition to supporting research and pedagogical projects, these practitioners also provided full-package usability testing services to clients at a cost. I chose to gauge the perspectives from these practitioners in a setting that is considerably different from the previous interviews because I wanted to know if there was a distinction in how design thinking and social innovation were conceptualized in an environment that can be influenced by academic thinking. I interviewed the two practitioners together. Both Kevin and Annie held the UX analyst title in the university usability services center. Kevin worked full time running the center while Annie only worked half time as she was completing a PhD at the university.

Kevin and Annie shared an understanding of design thinking in which they called the "logic model" of design thinking. Unlike inductive and deductive ways of problem solving, Kevin considered design thinking to be "abductive" reasoning. "That means understanding the effects or end results without necessarily knowing the causes, which are malleable," Kevin said. According to him, this model of thinking is common in UX decision making. Annie added that it is important to not reduce design thinking to radical empathy. Design thinking is in itself a tool that gets things done. Kevin also saw design thinking as an analysis tool, not just a design methodology. "Design thinking helps us identify what's at the core of a problem," Kevin stated.

From her work at the center, Annie has also realized that design thinking cannot be contained as a single event, but rather "more on-the-fly," in her words. Design thinking is a habit of mind for practitioners to always remember to put the holistic experience of the user in the everyday context of use. Kevin pointed out this mindset achieves best results when it is also shared with clients or collaborators. In cases where the power relations between the multiple stakeholders are intricate—such as when a product serves university administrators, instructors, and students—UX designers should advocate for students as they have the least autonomy in the design process. While Annie and Kevin didn't consider their work to be social innovation, their advocacy priorities aligned with the user advocacy concerns in social innovation.

A Call to Leadership: Social Innovations In and Out of the Classroom

In sum, the industry practitioners who provided their insights here all saw the values of design thinking in technical communication, including IT, medical design, documentation, and UX. It was clear in their responses that design thinking is first a mindset and methodology second. The way IDEO's Tim Brown and Jocelyn Wyatt defined design thinking is of interest here: "Design thinking—inherently optimistic, constructive, and experiential—addresses the needs of the people who will consume a product or service and the infrastructure that enables it" (2010, n.p.). Maintaining a design thinking mindset not only enhances product quality but also helps companies advocate for users. The technical communication practitioners who shared their perspectives toward user advocacy have highlighted the importance of empathy methods emphasized in design thinking. The activities that gear toward cultivating as well as exercising empathy in the design process are crucial to the later development of any product. The iterative nature of the design thinking methodology transforms the collaborative workflow, as Martin shared, it benefits the design teams and has staying power in the profession.

While none of the practitioners had directly related their work to social innovation, they all made implicit connections between their roles as technical communicators and the kind of innovative disruption that social innovation calls for.
As illustrated in Figure 3.2, when technical communicators practice design thinking in their innovation process, they are already engaging social concerns. Indeed,
social innovation may seem like an organizational mission, the acts of advocacy
can start with the individual practitioners. Their own sense of empathy toward
users and urgency in pushing user-centered agendas in design cycles can sow
seeds of social justice in their respective innovations. In the long run, these seeds
may germinate and grow into organizational priorities. As UX researcher Doug-
las Walls (2016) put it:

> Creating connections that support the lived experiences of underserved
> populations has immediate impacts. . . underserved populations find their
> needs not met in digital and offline contexts frequently because profit
> motives are not in place. Social justice focused UX work can address that
> market gap because success models should not be focused only on product
> revenue generation.
>
> (p. 2)

To empower current practitioners and aspiring professionals in engaging
social innovation, technical communication programs should include components for learning design thinking as well as leadership development. Through
design thinking activities, technical communication students should practice

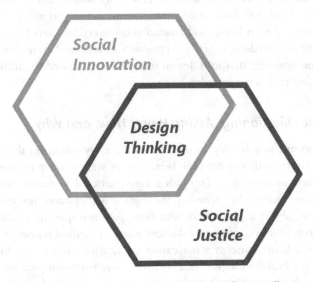

FIGURE 3.2   Design thinking provides a methodology for socially just and ethical
innovation

developing and deploying effective solutions that address social issues. In exercising so, students should learn to establish confidence in identifying talents and resources, and to convene productive collaboration teams for the purpose of social innovation.

Out in industry, we need to expand the professional agenda of social justice advocacy as part of the professionalization of technical communicators. Practitioners should be given opportunities to get involved in social advocacy projects and be rewarded for their investment. Programs like the 20% Project (popularized by Google)—where company employees are allocated 20 percent of their paid work time to pursue personally meaningful projects—can be adopted to promote social engagement by professionals. Corporate organizations should also provide appropriate mentorship to their aspiring social innovators by acquiring internal or external expertise to support employee projects. Since complex cannot be understood, let alone solved, by any single individual, strategic partnership and cross-disciplinary collaboration should be promoted and exercised between organizations in order to actualize social innovation. In the next section, I provide a few design thinking practices that can support user advocacy.

User Advocacy Methods Through Design Thinking

As a user-centered approach to problem solving, user advocacy is at the heart of design thinking. As the previous chapters establish, design thinking facilitates innovation through built solutions—via making—and entrepreneurial attitudes. The very first step in design thinking is empathy, which requires technical communicators and designers to set aside their worldly assumptions in order to gain true insights about particular users. This means actively resisting the urge to rely on previous design habits and instead enlist everyday users for their honest input throughout the design cycles. To practice so, technical communicators may employ some signature methods design thinkers have devised to cultivate empathy. Here I feature a few examples.

Journalistic Questioning: Asking What, How, and Why

A straightforward way to step into a user's shoe is by gathering the underlying factors and motives that drive user behaviors as well as their personal experience with certain scenarios. Through a conversational interview, the designer can gauge user feedback by recording the user's actual opinion, not the designer's interpretation, about a product or situation. For example, to empathize with patients who are often confused by doctors' notes or medical reports, a document designer can ask the patients what causes them frustration or confusion, how they peruse medical documents, and why they have particular preferences. These authentic feedbacks are more valuable than guesses made by designers who may have not been through the emotional experiences of an actual patient.

Photovoice

Thanks to the growing affordability of video and photographing technologies—via smartphones and hands-free devices—photo and video-based user studies have become popular empathizing methods in design thinking. Similar to diary studies, photovoice is a user-engaged method that gives designers a glimpse of their users' everyday lives. Participating users document the ways they interact with given scenarios using photographed images or video recording, hence photovoice. Designers can draw from these artifacts when interviewing their users and observe any emergent user behaviors in the documentary. This visual research method is different from other traditional user research approaches as it exposes the mundane yet insightful aspects of user experience that often get overlooked by designers.

Empathy Mapping

Much of empathizing with users is about getting a fuller picture about their *human* experience. Empathy mapping is a fundamental tool for capturing the affective characteristics of a user that may be used to create personas and user requirements later in the design process. The key in this practice is to consider the user as a whole person, not just an agent or character in a storyline. The traditional empathy map consists of four quadrants, each to capture respectively what a user says, thinks, does, and feels. For example, in the scenario where a user is using a grocery delivery service, the Says quadrant would be populated by what the user expresses out loud during an interview or usability testing session, like "I want to see my grocery options," "I like the price comparison feature," or "I don't know how to pay with this." The Thinks quadrant should contain intended as well as implicit expressions made by the user, such as "I believe every trip should be completed under an hour" or "I get annoyed when the deliverer picks up the wrong brand." The Does quadrant documents the actions and behaviors of the user during their interaction with the product, e.g., the user goes to the cart page to see the total amount spent on their groceries, or the user couldn't remove a selected item. Finally, the Feels quadrant encloses the user's emotional

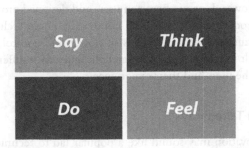

FIGURE 3.3   A simple empathy map for capturing human experience

state throughout the session. The content in the quadrant can overlap with the Thinks quadrant but focuses more on specific pains and gains, like confusion, satisfaction, anger, confidence, disappointment, etc.

Journey Mapping

Many designers confuse empathy mapping with journey mapping. While the former seeks to capture the user's whole internal experience in a given scenario, the latter concentrates on the external flow during an interactive process, hence *journey* mapping. The main purpose of journey mapping is to visualize a user's step-by-step travel through a scenario (like ordering an Uber ride), and capture the key stages of the process (e.g., setting up an account, entering payment information, entering a destination, choosing ride options, waiting for a ride, getting into a ride, interacting with the driver, arriving at the destination, rating the ride). Across these stages, designers would imagine potential high (happy) and low (dissatisfied) points so they can create interventions to help mitigate negative experiences and enhance positive ones.

Bodystorming

A word-play on "brainstorming," bodystorming is a physical activity where participants perform instructed actions using mock or actual products. This exercise aims to simulate scenarios as close to the reality experienced by users as possible. For instance, it's been reported by Apple Inc employees that when Steve Jobs was in the design phase of the revolutionary iPhone, he had several employees carry a wood block in their pocket and pretend to use it to make calls or text messages. This way, Jobs was able to observe user behaviors in a more natural state (albeit still within the workplace) and identify problems the users might face with his design. This method requires careful planning to create a real-life environment for participants to interact with a product. The results from this kind of study can be rich and insightful as it involves their full, embodied experiences.

These empathy-building exercises are just the starting point. User advocacy through design thinking requires the cultivation of a mindset that always pays attention to the power dynamic between users and designed products. Throughout the design process, empathy should continually guide technical communicators in recognizing the contested sites of power and control, and how these tensions may result in frustrations between colleagues with different agendas and ultimately affect the users and their experience.

Summary and Takeaways

While social innovation may sound like a popular fad to technical communicators, its exigency aligns with that of the social justice paradigm that is becoming increasingly dominant in technical communication scholarship and practice today. By situating user advocacy and social issues at the core of technical communication, this chapter presented design thinking as a methodology that supports social innovation, with the argument that technical communicators should pursue leadership in social advocacy. The case examples featured in this chapter demonstrated how design thinking can facilitate user-centered solutions. Through interviews with industry practitioners, I presented insights about these experts' perspectives on social innovation and the role of design thinking in their respective scopes of work. Altogether, these instances highlight the implications of design thinking and social innovation for technical communication, especially in programmatic innovation for the workplace and the classroom. Key takeaways from this chapter are:

- Technical communication is rooted in user advocacy and social justice. Design thinking can help level up technical communication's social advocacy through social innovation.
- Existing projects in technical communication and adjacent fields showed the importance of design thinking principles, i.e., participatory design, rapid prototyping, and empathy, in facilitating social innovation.
- Industry practitioners acknowledged the importance of design thinking in user advocacy but did not always recognize their work in terms of social innovation.
- Corporate organizations can lead the change toward more involved social advocacy by incentivizing technical communicators who pursue social innovation.

Learning Activity: Facilitate a Community Workshop

A productive way to uncover community needs and identify design opportunities is by conducting a community workshop. The community workshop method lets designers gather firsthand accounts of problems and user needs by interacting directly with those who are affected by the problem.

One crucial step to take in this research process is critical reflection *prior* to data collection. Before you design the workshop and approach your target community, you should first consider your own role and positionality as the design thinker in this project: What power and privilege (Walton et al., 2019) do you enact? How might you be transparent about these identities and let your community partners/participants learn about you? These questions can help designers and researchers see their own biases and goals in conjunction with the research agenda. Having answered these questions—and having considered the ethics involved in this research—you may decide to plan a community workshop with students and community partners to gather insights about a specific social problem facing the community.

In terms of logistics, you will need a space/room that can hold 15–20 people comfortably, walls or tables that allow participatory exercises, and sketching utensils like pens and sticky notes. This workshop may be led by students as a class, with 6–12 community members, for about 60–90 minutes. Below I recommend a simple flow for the workshop:

1. Introduce the purpose of the workshop. Students may do a quick (5 minutes) presentation about current challenges in assisted home living for senior citizens, and share their initial findings.
2. Gather informed consent. Let the community members know of their rights as participants of the workshop.
3. Do a breakout discussion. Organize community members into smaller groups (of 4–5) with each group being facilitated by 1–2 students. Students should present prompts that help community members think about their everyday experiences, wishes and hopes, struggles and pain points.
4. Do a large-group discussion. Students from each group report back the insights gathered from the breakout session, and have the community members identify the top three concerns they would like to see addressed. Students may write these concerns on a board/wall/paper on the table.
5. Conduct a "Dream Session" where all community members write their aspirations and suggestions for addressing the selected concerns on post-its, and put them to the board/wall/paper. Use arrows and lines to make connections between the post-its.
6. Have a student leader or two provide a summary based on the post-it notes by the end of the Dream Session.
7. Create a point-of-view (POV) statement using the user-requirement structure and ask the community members for feedback.
8. As you're concluding the workshop, find two or three community members and conduct brief one-on-one conversations to gather some information for creating some user personas. Capture actual quotes (verbatim) from these community members for later use.
9. Thank the community members for their time and participation. Provide any rewards/incentives or gifts as appreciation.

Reflection questions for workshop facilitators:

- When did participants seem particularly engaged? What were they doing? What were you doing?
- When did participants seem to have difficulty or were disengaged?
- What insights do you find most interesting?
- What is your revised POV statement?
- How can you use this POV statement to ideate possible solutions?

Notes

1. Johnson keeps the blog *I'd Rather Be Writing* (https://idratherbewriting.com/), writing regularly about technical communication trends and news, API documentation, information usability, among other topics.
2. The User Experience Professionals Association (UXPA) also maintains a "code of professional conduct" specifically for usability practitioners (last updated in September 2005), retrievable from https://uxpa.org/wp-content/uploads/2018/08/CoC_English.pdf

References

Björling, E., & Rose, E. (2019). Participatory research principles in human–centered design: Engaging teens in the co-design of a social robot. *Multimodal Technologies and Interaction, 3*(1). Retrieved from https://doi.org/10.3390/mti3010008

Brin, S. (2013). Why Google Glass? *TED Talks.* Retrieved from www.ted.com/talks/sergey_brin_why_google_glass?language=en

Brown, T., & Wyatt, J. (2010). Design thinking for social innovation. *Stanford Social Innovation Review.* Retrieved from https://ssir.org/articles/entry/design_thinking_for_social_innovation#

Buolamwini, J. (2016). How I'm fighting bias in algorithms. *TEDxBeaconStreet.* Retrieved from www.ted.com/talks/joy_buolamwini_how_i_m_fighting_bias_in_algorithms?referrer=playlist-the_inherent_bias_in_our_techn#t-16045

Dragga, S., & Voss, D. (2001). Cruel pies: The inhumanity of technical illustrations. *Technical Communication, 48*(3), 265–274.

Ethical principles. (1998). Society for technical communication. Retrieved from www.stc.org/about-stc/ethical-principles/

Haas, A. M., & Eble, M. F. (Ed.). (2018). *Key theoretical frameworks: Teaching technical communication in the twenty-first century.* Logan, UT: Utah State University Press.

inControl: A video game controller designed to be used with one hand. (n.d.). Northwestern University Segal Design Institute. Retrieved from https://design.northwestern.edu/programs/take-design-course/design-thinking-communication/projects/profiles/incontrol.html

Johnson, T. (2015). Innovation in technical communication. Presentation given at TCWorld India. Retrieved from https://idratherbewriting.com/innovation/#/

Johnson-Eilola, J., & Selber, S. (2013). *Solving problems in technical communication.* Chicago, IL: University of Chicago Press.

Jones, N. N., Moore, K. R., & Walton, R. (2016). Disrupting the past to disrupt the future: An antenarrative of technical communication. *Technical Communication Quarterly, 25*(4), 211–229.

Katz, S. B. (1992). The ethic of expediency: Classical rhetoric, technology, and the Holocaust. *College English, 54*(3), 255–275.

Medina, C., & Pimentel, O. (Eds.). (2018). *Racial shorthand: Coded discrimination contested in social media.* Logan, UT: Computers and Composition Digital Press/Utah State University Press. Retrieved from https://ccdigitalpress.org/book/shorthand/

Newbold, C. (2020). COVID-19 (re)opening decision trees: CDC visual guides to opening businesses, schools, childcare, and mass transit. Retrieved from https://thevisualcommunicationguy.com/2020/05/27/covid-19-reopening-decision-trees-cdc-visual-guides-to-opening-businesses-schools-childcare-and-mass-transit/

Noble, S. (2018). *Algorithm of oppression: How search engines reinforce racism.* New York, NY: New York University Press.

Norman, D. (2014). Living with complexity. *YouTube.* Retrieved from www.youtube.com/watch?v=Tj96KyC9zdI

Pfister, D. (2019). Technoliberal rhetoric, civic attention, and common sensation in Sergey Brin's "Why Google Glass?" *Quarterly Journal of Speech, 105*(2), 182–203.

Project EMAR. (n.d.). MeLab. University of Washington. Retrieved from http://depts.washington.edu/melab/projects/project-emar/

Rittel, H., & Webber, M. (1973). Dilemmas in a general theory of planning. *Policy Sciences, 4*(2), 155–169.

Rose, E., & Björling, E. (2017). Designing for engagement: Using participatory design to develop a social robot to measure teen stress. In E. Keller (Ed.), *Proceedings of SIGDOC'17* (Paper no. 7, pp. 1–10). New York, NY: ACM.

Salvo, M. (2009). Ethics of engagement: User-centered design and rhetorical methodology. *Technical Communication Quarterly, 10*(3), 273–290.

Soule, S., Malhotra, N., & Clavier, B. (n.d.). Defining social innovation. Stanford Center for Social Innovation. Retrieved from www.gsb.stanford.edu/faculty-research/centers-initiatives/csi/defining-social-innovation

Usability.gov. (n.d.). Benefits of user-centered design. Retrieved from www.usability.gov/what-and-why/benefits-of-ucd.html

Walls, D. (2016). User experience in social justice contexts. In S. Gunning (Ed.), *Proceedings of SIGDOC'16* (Paper no. 9, pp. 1–6). New York, NY: ACM.

Walton, R., Moore, K. R., & Jones, N. N. (2019). *Technical Communication after the social justice turn: Building coalitions for action.* New York, NY: Routledge.

4

MAKING AND DESIGN THINKING
AS PEDAGOGICAL STRATEGIES
FOR SOCIAL ADVOCACY

Overview: This chapter situates design thinking and making in technical communication pedagogy. Following the exigence underscored by the previous chapters, this chapter showcases the development and deployment of design thinking and making in a technical communication service course. The results of this pedagogical case study support the contention that design thinking and critical making help technical communication studies locate, define, and understand complex problems; it allows them to practice empathizing with users, ideating and prototyping solutions, and implementing iterative design. The chapter leaves readers with a set of pedagogical exercises in design thinking, such as empathy mapping, contextual inquiry, and rapid prototyping, which can be modified or scaled to accommodate different class sizes and degree levels.

Making in Technical Communication Pedagogy

In Chapter 2, I discussed the impact of the maker culture and how making as a learning exercise can teach students to solve problems by tinkering with prototyping and digital fabrication technologies even if they do not have prior experience with the tools or specific manufacturing processes. Through the design thinking methodology and mindset, I posited such activities as a step toward user advocacy, which can promote critical awareness of social issues. I have featured some case examples and industry perspectives to maintain this thesis in Chapter 3.

The availability of easy-to-use digital fabrication tools like 3D modeling and printers, computer numerically controlled (CNC) milling machines, and water jets and laser cutters allow users to perform lower-stakes experimentation. Through trial and error, students can learn by adjusting their problem-solving process—changing their measures, adapting from previous conditions, modifying assessment criteria, etc. The ability to tinker with tools and adapt to problem situations are increasingly desirable workplace competencies that technical communication courses across the board can help cultivate and strengthen. Technical communication courses should afford students with access to the tools and technologies they need to experiment and prototype, moving ideas from conceptualization to materialization.

Further, since making does not happen out of context, this critical-creative tinkering approach to learning (see Koupf, 2017) re-envisions the purpose of the classroom and the roles student makers could play in affecting social change. Through design challenges, students participate in social innovation exercises and community initiatives that can make a positive difference to those of whom the projects benefit. Making and design thinking are, thus, more than just a skills-based development, but a pedagogy of social justice that motivates students to become active advocates in their respective capacities.

To mobilize this approach to learning, I have experimented with using design thinking as a pedagogical framework for a technical communication course. My goal was to explore the viability of design thinking methods and processes in teaching technical communication genres, workflow, and learning outcomes pertaining to the profession's needs. Although design thinking is no stranger to those who practice or teach user experience design and project management, both of which dominant sub-areas of technical communication, only a handful of technical communication instructors have applied design thinking values and methodologies in general technical communication courses, such as document design, grant writing, publication management, content management, and technical editing.

Through my own exploration, I have intended to demonstrate a possibility of integrating design thinking as the core of a general technical communication course. Given the commonality (and necessity) of the introductory course in most technical communication programs in the United States, I have focused on re-envisioning the design of an elemental technical communication course at a research university—WRIT 3562W Technical & Professional Writing—which, at the time of this writing, also served as a "service course" for undergraduate studies at the university. More details about this course and my redesign process are provided in the coming section. I will, however, leave the discussion about the nature of a "service" versus "non-service" technical communication course for another occasion due to the complexity of such discussion. Here, I focus instead on presenting the rationale for creating a design-centric technical communication course that helps students develop the competencies highlighted in the opening of this chapter.

In the following sections, I discuss such rationale at the intersection of design thinking, making, and technical communication objectives. By providing the design and deployment details from the pedagogy case study, I aim to demonstrate how design thinking can support technical communication pedagogical goals. The findings from the case study show the affordances as well as limitations of design thinking in achieving technical communication learning outcomes. I end this chapter with exercises and prompts that technical communication instructors may adopt and adapt to experiment with design thinking in their own courses.

A Rationale for Making and Design Thinking

Design thinking and making, both seemingly non-academic ideas and practices, are yet excellent processes to apply in technical communication courses to (re) invigorate the learning experience in these courses. They promise to motivate students to empathize with users, create and test radical ideas, and leverage collective knowledge to address complex social problems. I have located the value of these processes by observing amateur makers build things in makerspaces and speaking to them about their perception of such approaches to learning. As described in Chapter 2, I conducted site observations at three leading academic makerspaces prior to designing and deploying a design thinking-powered technical communication course as part of this study. From these makerspace observations, I learned that students found design thinking and making a natural way to apply their content knowledge to solve the problems at hand. In fact, design thinking had manifested as a habit of mind for these students. Design thinking terminologies were ingrained in these students' vocabularies; they shared in the interviews with me how they "designed prototypes," "practiced iterative cycles to invention," and "devised human-centered solutions."

Another key motivation for integrating design thinking with technical communication pedagogy was its tendency for meaningful collaboration, as in peer-to-peer learning. All of the students I interviewed during my makerspace observations referenced the positive social experiences in their respective makerspaces and how much they learned from other makers in the space. Similarly, most of them talked about helping others in the makerspace as well. This realization is ideal for promoting situated learning in an informal setting; it can also be an important stepping stone for acquiring technical communication skills, since the content and technologies are always evolving. Students may learn specific methods and problem-solving strategies by observing others' approaches, as well as through their sharing of expertise. Thus, when creating my design thinking technical communication course, I incorporated team activities to actualize such potential of collaborative learning.

Designing Design Thinking in Technical Communication Pedagogy

Central to the technical and professional writing course that I chose to redesign was the well-established assignment sequence it offered to introduce students to the popular genres of technical communication. They included, in the following order:

- Memos
- Technical definition & description
- Instructions set (including technical graphics)
- Mini usability test & report
- Technical proposal (including budget & timeline)
- Technical or feasibility report
- Presentation

The goal of this course was to let students practice writing in mainly non-academic contexts, focusing on highly technical topics (e.g., "Evaluating a neighborhood's readiness for hydroponics gardening" and "Determining the feasibility of hybrid buses at a small liberal arts college"). Between 2014 and 2019, more than 20 sections of this course (with 24 students in each section) were offered every spring and fall semester. Due to its "service" nature, these course sections were made up of technical communication majors as well as students from business majors, architecture, agricultural science, nursing, and dentistry, among others. Also, because this course had a special writing-intensive designation ("Definition of a writing intensive course," 2010) by the university's writing board, instructors were asked to follow the pre-approved assignment sequence as closely as possible.

At first glance, restructuring the original assignment sequence was deemed a daunting task. Having taught this sequence twice—once onsite (face-to-face; synchronous) and once online (asynchronous)—I have experienced the intensity of this course and the amount of work students produce through the predetermined assignments. When approaching the redesign effort, I consulted with the director of undergraduate studies as well as the department chair; both happened to be members of my dissertation committee. After a few rounds of discussion, my committee members and I saw a possible revamp of the original course structure without having to remove any of the key assignments from the course. Since technical communication and design thinking are essentially problem-solving activities, I was able to frame the redesign effort around the notion of finding, making, and presenting solutions to complex issues. I used the language of design thinking and offered a "design challenge" approach to this technical communication course.

By mapping the assignment sequence onto the design thinking process (see Figure 4.1, expanded from the Stanford d.school basic model, with an additional "implement" phase), I presented the key assignments in this course as primary elements of the problem-solving process. Students were assigned into teams they kept throughout the semester. Progressing through the iterative design process, students completed the major assignments in a new sequence and were still able to experience the original intentions of the course.

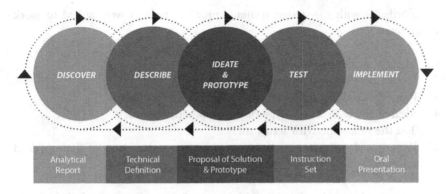

| Analytical Report | Technical Definition | Proposal of Solution & Prototype | Instruction Set | Oral Presentation |

FIGURE 4.1   The assignment sequence in WRIT 3562W mapped onto the design thinking process

The "design challenge" pulled together the new assignment sequence under a common theme. To introduce students to the uses and purposes of each technical genre in the assignments, a design challenge was integrated to provide contexts for practice. A signature activity in design thinking bootcamps, a design challenge typically revolves around "wicked problems" (those lacking well-defined single solutions) and requires the challengers to work in cross-functional teams and exercise the design thinking process:

1. Empathize with users & stakeholders
2. Define scope of project
3. Ideate radical solutions
4. Create prototypes
5. Test prototypes
6. Iterate design
7. Present or implement solution

In my redesigned technical and professional writing course, students were challenged to enhance campus experience by addressing a specific wicked problem pertaining to student life on campus. The prompt for the design challenge read:

> *Your team will learn about the experience of students in the UMN campus community and identify a potential problem they face in a specific domain of the campus experience. You will define this potential problem and ideate a viable solution to address the problem. You will create a prototype for your proposed solution, which you will use to test with actual users. Finally, you will present your idea with details on the costs and benefits for implementing your proposed solution in context.*

In accordance with the design thinking process, students were tasked to work with a team to:

1. Identify a campus experience problem.
2. Define the potential problem.
3. Propose a viable solution to address the defined problem.
4. Ideate and create a prototype of the proposed solution.
5. Test the prototype with actual users.
6. Present a plan for implementation with costs and benefits of the proposed solution.

Ideas presented to students as potential domains for investigation included student housing and meal plans, transportation, campus safety, and learning resources. Students were also invited to consider other ongoing issues around campus climate, extracurricular activities, and health services.

As shown in Figure 4.1, the major assignments corresponded with the design thinking process so students could complete their design challenge as in Table 4.1.

Unlike the previous assignment sequence in WRIT 3562W, where the analytical report was assigned as the final research project, I modified it so that it came as the first major assignment and students could begin locating the potential problem areas and specific issues faced by the campus community. Students began by brainstorming ideas in their assigned teams during the first week. I also conducted a design thinking orientation to introduce students to the design thinking process, ways to think radically about problems, and potential approaches to addressing them. Table 4.2 shows the complete description and weight for each assignment.

To encourage students to look beyond the administrative or logistical aspects of the problems they have identified, I emphasized the importance of focusing on

TABLE 4.1 The design thinking process and correlating assignments and goals in each phase

Design thinking phase	Assignment	Goals
Discover/empathize	Analytical report	Learn about people and the context of their problem
Describe/define	Technical definition and description	Synthesize learning from stakeholders/users
Ideate & prototype	Proposal of solution and prototyping	Iterative ideation and materialize designed solution
Test	Instruction set	Test ideas and prototypes with actual users
Implement/present	Oral presentation	Pitch chosen/refined solution to stakeholders

TABLE 4.2 Major assignment descriptions and weight in percentage

Assignment	Description	Weight
Analytical report (3 weeks)	Students will identify a problem on campus that could be addressed with existing/emerging technologies or technology-enhanced processes. Through observation, analysis, and data collection (such as qualitative interview, survey, and content analysis), students work in teams of three to identify a wicked problem within the campus community, determine researchable questions, and ideate ways to address their research questions. The goal of this 1000-word report is not to solve the problem, per se, but to initiate a plan for a semester-long multimodal project.	15%
Technical definition and description (2 weeks)	In a 500-word memo, each student team selects a technical term pertaining to their design project and provides a concise definition of the specialized term. The definition should be accompanied by a detailed explanation of objects, places, or processes as the description of the technical term.	10%
Proposal of solution and prototyping (6 weeks)	Each student team proposes a solution to the problem and/ or research question they have identified in the analytical report. This 1000-word proposal of solution should be written with a specific audience in mind. The proposed solution must be prototyped either in a digital or physical form. The prototype must be turned in to the instructor and will be presented to the class at the end of semester.	25%
Instruction set (4 weeks)	Each student team will organize and write an instructional procedure to enable a specific audience for the proposed solution of the identified problem. The instructions set must have at least 20 steps, include at least 5 visuals/ illustrations, list the materials required, and include a warning/caution step. This set of instructions will be tested on by at least two users. The final instructions set should reflect revision based on the results of usability tests.	15%
Presentation (2 weeks)	Each student team will organize and deliver a 15-minute professional presentation about their identified problems, design/prototyping processes, proposed solutions, and final prototype.	10%
Reflections (1 week)	Each student produces a 500-word reflection narrative about their learning experience with the assignment sequence and the semester overall.	5%

the *experience* of other students within the problem, such as the experience of dining on campus or using the shuttle systems, rather than on the *personnel*, such as the university administration or a professor. The specific problem should also deal with *technological* issues; that is, it needed to be a problem that could be addressed with changes (or addition) to its existing technological design. Together, these emphases prompted students to consider the social dimension of their wicked problems, and to pay attention to issues of equity, accessibility, and technological literacy gaps, among others.

Students were also made aware of the requirement to devise and create a tangible prototype for their proposed solutions, and, thus, they must consider the available resources for prototyping. Fortunately, I was able to secure programmatic support from the IT department of my college unit, where a digital fabrication specialist, a graduate student assistant from the archeology department, was available during the semester to provide technical support to my students. Students were given quick demos and tutorials on tools like 3D design and printing, 360° imaging and video recording, and soundscape design and mapping, as well as emerging technologies such as virtual and augmented reality headsets (HTC Vive, Oculus Rift, Google Daydream) and wearable gadgets (Pebble, Apple Watch, Google Glass).

I recognize and acknowledge the privilege of access to these tools and technologies at a large research institution. Many instructors with whom I have shared this pedagogical direction have voiced concerns about technological and technical support from their own institutions, and I have shared that these are not the primary focus of a design thinking centered technical communication pedagogy. While the tools provided students with better affordances in invention and delivery, they were, nonetheless, just tools. As I report later, most students from my course chose to use lower fidelity technologies to address their problems and create prototypes, i.e., wooden sticks, craft papers, and other up-cycled objects, as well as screenshots, digital wireframes, and dummy interfaces. The sophisticated technologies that were presented to students required more investment in terms of time (for learning the tool) and commitment to incorporating it in their final product. Instead, many students concentrated on their conceptual design and delivered lo-fi prototypes that captured their ideas, minus the flashiness. In retrospect, this was a more desirable outcome from a pedagogical viewpoint since the point of the design challenge was to inspire creative-critical problem solving, not building technology skills, per se.

Datapoints

Twenty-four students were enrolled in this case study, and all of them remained in the course throughout the study. These students represented five colleges of the university. Most of them came from the College of Food, Agriculture, and Natural Resource Sciences (CFANS) and the College of Liberal Arts (CLA). The majority of the students were juniors (17 of them), but one of them was a sophomore, and six were seniors. I used multiple methods to assess the impact of design thinking on students in this course including project evaluations, qualitative interviews, and autoethnography (via my teaching notes). Throughout the semester, I kept a running log of my teaching notes to document emergent ideas and observations. I wrote about students' reactions to the design challenge and assignment sequence, as well as my own attitude toward teaching design in the course. At the end of the semester, four students participated in qualitative interviews about their experiences with this course. To triangulate the data collected, I evaluated all final design projects completed in this course to determine the impact of the design thinking approach in meeting the student learning outcomes.

Cultivating Empathy

I have a vivid memory of the immediate reactions from students when I introduced the design challenge to them during the first week of the course. A few of the senior students received the instruction calmly like it was just another team project to be completed in a typical class. Some students looked particularly interested in the opportunity to produce a 3D prototype—an unusual assignment for a "writing" course. Others were unsure if they would be able to pull off completing the challenge as assigned. A few students came up to me after the first class meeting to query about my expectations for the final product, attempting to gauge the scope of work and commitment necessary for the design challenge. None of the students, however, asked to be exempted from the design challenge, nor did anyone request a different project.

The first step in the design challenge was to develop empathy for stakeholders (primarily students within the campus community) before determining the specific scope of the project. In teams of three and four, students spent a week going around the campus to observe how the students interacted with various facilities and spaces on campus. They conducted contextual inquiry (see exercise prompt at the end of this chapter) by way of informal interviews, such as asking students in a cafeteria how they use their meal plans and talking to friends in the dorm rooms about housing options on campus. These findings helped teams locate problematic areas that they felt capable of addressing as part of the design challenge. Using empathy mapping (see pedagogical exercises section), teams created user stories and requirements to guide their project direction.

A notable distinction between this initial phase in the design challenge compared to conventional academic research projects (which often began with the researcher's point of view on particular problems) was the emphasis on empathy, which led to user-centeredness in problem solving. It encouraged students to move from a designer/researcher-centric approach to problems to a user-focused practice. Guided by empathy and user stories, students learned to involve those who were affected by the problem early and always in the design process.

Defining Project Parameters

By the end of the third week of the design challenge, each team prepared and submitted an analytical report that described the particular problem area on which they would focus, the target audience of their potential solutions, the exigence for action, the methods of intervention, and the projected expenses and limitations of the design challenge. This allowed me, as the instructor, to get an overview of each team's focus and provide guidance on their methods and design plan.

While none of the reports in this instance required any major revision, some did have to narrow their focus. Some teams showed a huge ambition to tackle long standing problems (like the lack of parking spaces on campus) with few considerations of the limited resources they might have in this project. Thus, in my review I provided suggestions to reduce the project parameters for these teams, recommending they focus on more achievable deliverables. For example, instead of attempting to address the issues with availability of street parking and ticket prices, a team could, instead, look at new carpooling options or rethink how lots display their available spaces through digital means.

To help teams dive deeper into the problem areas they were choosing, the technical definitions and descriptions assignment let students select a technical term pertaining to their design project and provide a concise definition of the specialized term. The definition should be accompanied by a detailed explanation of objects, places, or processes as the description of the technical term. This assignment prompted students to conduct preliminary research that helped them identify relevant studies that have been done within their chosen area. It also helped them find their niche in the design challenge based on the understanding of existing research.

At this juncture, teams created some initial deliverables for the design challenge. Having students articulate their projected deliverables and ways to accomplish them early in the design challenge helped create the criteria for later use in evaluating the success of their projects. The sets of deliverables were made into a backlog of to-do items, each with an assigned start and completion date. This backlog became the checklist that kept teams moving along the design process, as well as a springboard to their ideation exercises.

Ideating Together

One of the key differences the design challenge assignment sequence offers technical communication pedagogy is the emphasis on collaboration. As students worked as a team to complete not one but all of the required assignments in the design challenge, they learned to communicate personal needs or concerns, resolve conflicts, and share tasks. Students practiced prioritizing goals and distributing workload. Design thinking favors a collaborative approach to problem solving as it promotes greater creativity and equity in designed solutions.

The ideation phase began when teams exercised their imagination to create multiple solutions to address the problem specified in their analytical report. This was a phase in which students both enjoyed and loathed during the design challenge due to the ambiguous nature of the ideation process. Many initial ideas were sketched and then scratched. Most teams produced documentation of their idea generation in their proposal assignment as a way of "showing the work," although it was not a required component. The collaborative nature of this process led to cross-disciplinary considerations in the ideated solutions. Students brought their developing disciplinary knowledge to the design table and used it to inform the selection of their chosen design direction.

This process adds value to technical communication pedagogy, especially for courses without a specific collaborative design component, as it creates the opportunity for students to assume authority in the creation of practical yet creative solutions to problems they deem important to address. At the same time, it challenges them to observe and consider the different expertise and traditions brought upon by the various disciplinary backgrounds of the team members. Design thinking provides a viable framework to facilitate this process by encouraging "radical" interactions within cross-functional teams.

Prototyping Testable Solutions

For students, perhaps the most exciting part of the design challenge was the prototyping phase. Again, the requirement of the design challenge project included the building of a tangible mockup (whether a 3D object or a wireframe interface) of the solution proposed by each student team. Throughout the course, students were introduced to various rapid-prototyping tools that were made available via the different units of our university. The advanced imaging lab offered 360° still and video recording technologies that students may use to create immersive simulations. The medical devices center and library makerspaces had 3D modeling applications and 3D printers that allowed students to fabricate plastic prototypes of their design. The emerging technologies lab collected several virtual reality and augmented reality headsets that students could use to test out their immersive simulations or create room-scale 3D paintings using applications like Google Tilt Brush.

Students were reminded not to let the technology drive their design directions. Rather, they should use these tools to materialize their ideas, and only if the outcome would match the intended design. It was also crucial to caution students to avoid spending excessive time figuring how to operate a tool. The goal of the design challenge project was to expose students to making/building/prototyping as a productive exercise in the design process, not to focus on learning a tool, per se. Ultimately, the purpose of prototyping was bringing to life the designed solutions so teams might use the prototypes in user testing sessions.

As part of this process, students created technical instructions for users to test the prototypes. Additionally, the instruction set was tested for clarity and accuracy during the user testing sessions. By the end of this phase, teams were given the opportunity to revise their prototype and instructions before organizing an oral presentation of their final solutions to a live audience.

Advocating for Change

Although the design challenge prompt did not necessarily steer students into a social justice direction, the resulting projects demonstrated awareness toward social responsiveness. Students designed and delivered solutions that put user equity at the core. Their overall approach to the specific problem at hand showed an increased motivation to advocate for user needs and combat unjust systems through radical designs.

As the design challenge project approached its due date, students planned a presentation of their solutions by providing a narrative of their problem selection, user analysis, design, testing, and iteration methodologies. At this point, students had learned to situate their proposed solutions in the social contexts of the problem areas they were addressing through the design challenge. From the literature that complemented the course, students recognized the need to not only critique social problems, but participate in the process of advocating for change. The design challenge project gave them a low-stakes platform for such practice. Teams articulated their ambition to affect social change through their innovation in the oral presentations.

From the focus group interviews, as I detail later, I learned that some teams had gone to present their proposed solutions to administrators who were in a greater position to potentially implement these solutions. This was not a requirement in the design challenge project. The fact that students felt compelled to extend their ideas to the authorities showed both a high level of ownership by these students towards their projects and the impact that design thinking has as a problem-based learning approach to technical communication.

Student Projects

By the end of the design challenge, I evaluated all student projects for their quality and effectiveness in addressing each of their specific problem areas. Three teams focused on campus parking problems, two teams on student dining, one on campus safety, one on campus housing, and one on campus navigation. For teams that were addressing similar problem areas, they concentrated on different issues. For instance, the students addressing campus parking problems focused on three related but distinctive issues: ticketing, parking space notification, and ride-sharing. Table 4.3 provides a summary of each project's problem area, proposed solution, and prototype.

All of the student projects met the requirements of the design challenge. They each completed the assignments that made up the sequence and presented tangible solutions with prototypes by the end of the project. Of the eight final prototypes, five were hi-fi wireframes with clickable buttons and multiple pages, two were website mockups with full content (including copy and images), and

TABLE 4.3 Overview of student design challenge projects and outcomes

	Problem area	Proposed solution	Prototype
Team 1	The lack of flexibility in meal plan spending options and students' unhealthy dining behaviors.	"To implement a points system in the dining hall using a device called PointPost. Each station in the dining hall is allotted a certain amount of points. Users have the option to view their point balance through an app called NextJEN PointPost. Students will have more control over their spending habits and will only grab food that they wish to eat."	*PointPost*—a scanning station for meal points.
Team 2	The lack of certain nutritional options in the university dining halls.	"The idea of the application is that students may forfeit a 'meal' from their meal plan in order to procure groceries and, in so doing, would have a small amount of their meal plan's cost credited back to them."	*Gopher Grub*—an application for tracking one's nutrients and reward them financially for logging their diet
Team 3	The lack of navigational tools offered to find one's way around our massive, 3-campus university.	"360 degree, interactive views of both indoor and outdoor pathways and areas are implemented within the interface, enabling the user to find physical markers within the building to aid in recognizing the space they are locating."	*MapIt*—an app with real-image mapping for indoor and outdoor navigation.
Team 4	Campus night-time safety concerns.	"The installation of campus Help-U with U-Travel interactive displays in strategic locations of university properties to effectively connect distressed students and other members of its community to a friendly, system that serves as accessible navigational service."	*Help-U*—a website to aide in building/campus navigation, monitoring, and dissemination of building information for University of Minnesota patrons. *U-Travel*—a centralized website for three University sponsored websites: Parking and Transportation Services, Public Safety and the U, and Safe-U. A digital display to feature Help-U and U-Travel.

(*Continued*)

TABLE 4.3 (Continued)

	Problem area	Proposed solution	Prototype
Team 5	Overpriced luxury student housing that limits students' off-campus housing options.	"An apartment complex with a focus on practicality and opportunity for student-subsidized rent."	*H.A.M. Student Housing*—a website for scheduling a tour, applying for a lease, applying for a job within the complex, tabs for overseeing rent subsidization, as well as pages that allow potential residents to view floor plans and read about the housing provider's mission.
Team 6	Expensive campus parking costs and penalties.	"Our group thought prototyping kiosks around parking ramps, lots, and garages would help university students avoid unnecessary payments."	*Tiki*—a digital ticket counter that sends parking tickets to user's mobile device.
Team 7	Commuting students face high costs for on-campus parking.	"Our proposed solution is to create a collaborative mobile application between Uber and University of Minnesota."	*M-Uber*—Just like Uber but with additional features that benefit University of Minnesota students.
Team 8	Difficulty in locating available parking spaces around campus.	"To improve experience with real-time viewing of parking spaces via color coordinated map, various maps covering East Bank, West Bank and the St. Paul campus, and cheap & convenient mobile payments."	*ParkSmart*—an app with real-time display of parking availability around three university campuses.

one was a 3D-printed prototype. Notably, students learned to communicate their project's purpose and goals more clearly as the design challenge progressed. This was reflected by the quality of writing in the team assignments. Most teams received an average grade (i.e., B) for the analytical report and technical definition and description assignments. Five of the eight teams received an excellent grade (A) for the instruction set assignment. All but one team received an A for the proposal and presentation assignments.

Student Responses

By the end of this course, I offered an open invitation to welcome students to share their experience with me in the form of an interview. One student volunteered to do an individual interview, and a team of three students signed up to complete a focus group session with me.

Hannah, who completed the individual interview with me, was an animal science major who took this course to fulfill her upper-division writing-intensive requirement in the major. During the interview, she spoke mainly of the design challenge as a whole and why she thought it was a valuable experience. First, Hannah noted the ambiguity in the challenge and how that inspired her and her team to approach problems that were outside her immediate experience:

> I enjoyed the ambiguity and creativity to pursue a project we wanted, within reasons. For the most parts I enjoyed working on the assignments. They were very intentionally designed to be team-based. It is hard to do everything by yourself, especially if you don't have the skills necessary to solve the problem.

When asked about her learning experience, Hannah said that she practiced applied problem solving through making an actual prototype to address the problems at hand. She also noted that she learned the importance of organizing a workflow so her team could stay on schedule:

> The most challenging assignment, for me, was the proposal and prototype assignment. Most of the other assignments were built around the prototype and how it works, how users would operate within it. But, we didn't have our prototype made until a few days before we had to turn in the proposal, so writing the proposal before having the prototype in hand was difficult. And, then the instruction set, too [needed the prototype]. As the proposal assignment was wrapping up, we were writing the instructions for using the prototype, but we wondered how were we going to do that without a full prototype.

Finally, Hannah shared a sentiment that I agreed to be a key distinction to learn from the design challenge:

> We have more than enough people in the world who *talk* about problems; we need more people who can *solve* them.

The most important takeaway from my interview with Hannah was getting to know how a student negotiated her way through a wicked problem that was unclear to her and her team at first. I learned from this individual interview that one key attitude to cultivate was embracing ambiguity.

In a separate session, I met with three students who were assigned to the same team for the design challenge. Two of these students were female (Sheryl and Shelby) and one was male (George). George was pursuing a bachelor of independent studies, while Sheryl was an economics and actuarial science major, and Shelby was a health management major.

At the interview, these students were asked to share what they found valuable in the course through the design challenge and what were some challenges they faced. Shelby said she learned to apply the design thinking methodology from an initial orientation to her team project, as well as seeing other teams' design process:

> In the beginning I wasn't quite sure what the design thinking activity was about, but as the semester went on, I was able to see what you were trying to get at. I also thought it was cool to be able to see what the other groups were doing, not really giving out exactly everything but a preview. So, I thought to myself, "that is so cool what can we do that is like theirs and how can we make ours different at the same time."

The three students also talked about how the team might benefit from sharing with one another their individual strengths and weaknesses in the beginning of the project. George noted the importance of not just giving one another tasks they think they were good at:

> If we only focused on our strengths, I don't think I would have developed areas that are my weaknesses.

Sheryl said that she learned how to write better from looking at George's writing. She also highlighted that she had learned more about technical writing from the design challenge project:

> Because of this course, I now have a deeper understanding of technical writing. I know how to identify a problem, describe it, gather information and data, and come up with conclusion about the problem or recommend solutions.

Sheryl added that her team learned to take the initiative to find resources that were related to their own project:

> I think it is also important for students to reach out to units and departments that are pertinent to their own projects because the instructor is not that magic, and he can't do everything.

From the team interview, I learned that students found the design challenge to be appropriate for practicing collaborative problem solving. It was encouraging to hear students reiterate the design thinking methodology in their responses, like understanding users and iterating design.

The results of this design challenge case study supported the contention that design thinking and critical making helped technical communication students locate, define, and understand wicked social problems. It allowed them to practice empathizing with users, ideating and prototyping solutions, and implementing iterative design. Through the evaluation of student projects and interview responses, I observed that students accomplish the learning outcomes designated for this course through active problem-solving exercises in the design challenge assignment sequence. As I show next, in the pedagogical activities, these exercises can be scaled and incorporated into technical communication courses of any size and degree level given their flexibility in methodological design and use.

Pedagogical Exercises

To help those who are interested in implementing a design challenge or similar activities in their technical communication courses, I offer here a set of pedagogical exercises in design thinking that can be modified or scaled to accommodate different class sizes and degree levels.

Contextual Inquiry

An analytic process to understand the behaviors, actions, inner workings, and cultures of users in-situ, contextual inquiry is usually done with methods like interviewing and site observations. technical communication students can benefit from contextual inquiry exercises as part of empathy building and problem definition stages of a design project.

> Purpose: To understand how new subscribers navigate your product's homepage and make content selections.
>
> Prompt: You are one of the user interface designers for Netflix, an online streaming service provider. You would like to know what new subscribers do on the web homepage of Netflix and how they choose their first content to watch.

Practice: Conduct a contextual interview to gather information about the experience, needs, and motivations of various new Netflix users. Listen for users' desires and actions that may be informed by prior experience, cultural influence, or mental models. Start by creating a set of criteria for selecting your interviewees (e.g., they must be first-time Netflix subscribers), then recruit a handful of participants, and conduct a few 30-minute individual or small group interviews with them. Moderate the conversations to focus on the participants' emotions, expectations, and contexts of using Netflix.

Point-of-View (POV) Statement

Once students have collected initial user feedback on particular issues, they may generate their POV statement, an exercise that can be done with users or by the design team alone. This exercise aims to articulate specific users based on the initial user feedback. The goal is to translate user experience into user stories.

Purpose: To create a set of user stories and user requirements to guide your design and evaluation of designed solutions.

Prompt: You are an interaction designer for Cactus, a new location-based RSS feed for theatre goers. Using the initial narrative data gathered from contextual inquiry, create two user stories and accompanying user requirements to help determine the design direction for the beta version of Cactus.

Practice: Use the following template to articulate a user story. If conducting this with users, let them fill out the blanks first.

As a _____ (user role)
I need _____ (user requirements)
So that I can _____ (user goal).

Radical Imagination

Ideation is one of the most exciting parts of design. To inspire creative solutions, radical imagination is a process to come up with "out of the box" solutions that can be materialized later in the design process. While radicality is subject to situations and context of use, this exercise encourages students to resist traditional constraints and come up with ideas that would shatter conventions and ideologies.

Purpose: To devise unconventional ideas in order to promote more ethical and equitable solutions to problems.

Prompt: You are charged with ideating an improved textbook model for Argo, an open textbook publisher that aims to deliver "free" textbooks to college students. Provide six radical solutions that would re-envision the future of textbook publishing.

Practice: Do not be constrained by financial, technical, technological, or human resources at this time of ideation. Spend as little time as possible to generate any many "wild" ideas. Use sketches, line-drawings, or other quick idea-to-paper methods. Once you have ideated about a dozen creative directions, combine related ideas into more tangible solutions. Present at least six radical ideas by the end of the exercise.

Rapid Prototyping

Design thinking prioritizes material solutions over conceptual/abstract ideas. Rapid prototyping is a way to actualize designed solutions by inviting students to build/make their ideas into tangible forms. Often, students would expect to create high-end or polished outcomes, but the goal of prototyping is to materialize design in usable conditions, which can usually be accomplished through low-fidelity (low-fi) prototyping. This exercise lets students practice creating physical mockups of their designed solutions without worrying about the external quality of their prototype.

Purpose: To materialize design ideas using readily accessible materials and tools so that it can be used for testing or to gather feedback before returning to building and iterating.

Prompt: Create a prototype of your selected design (you may choose from the radical imagination ideas above) using either computer assisted design software like Adobe InDesign or other freeware online, paper materials, storyboards, or other mockup tools. Your prototype does not need to be complete or even functional.

Practice: Select your idea for prototyping. Choose a prototyping method. Ideally, you should aim to complete your prototype within 2 hours. Do not focus on minute details in your prototype; create an object manifestation of your idea that gives the design idea a look and feel tangible enough to receive meaningful feedback. If your prototype doesn't come together initially, iterate your design as you build a revised prototype.

Summary and Takeaways

This chapter has detailed the phases student teams underwent to complete the design challenge. In essence, a design challenge is an active learning pedagogical approach that motivates students to work in teams to identify wicked problems, build and show empathy toward users, ideate and prototype radical solutions, and test and implement these solutions with the intention to cultivate positive change. Key observations from this chapter are:

- Collaborative problem-solving increases student engagement and can lead to more productive course learning experience.

- technical communication students learn to design user-centered solutions through empathy building and direct interactions with stakeholders.
- A design challenge strings together design thinking methodologies and foundational technical communication activities to introduce students to important technical communication genres.
- A socially situated design challenge orients students to user advocacy and radical imagination.

Learning Activity: A(nother) Design Challenge

In a team of 3–4 members, consider the following problem statement and devise a strategy for designing a viable solution to address the problem. In your strategy, be sure to specify the users and stakeholders involved in the situation, technical terms that require explanations, design methods, and a plan for implementing change.

> Every year, half a million people are injured or killed in traffic-related deaths due to texting while driving. Although there are laws in place to lower device use while driving, these scare tactics have not been effective in reducing such dangerous habits. Using available materials and tools in your classroom, design a strategic solution to address this wicked problem.

When you have finished sketching your strategy, discuss with your team members the following questions:

- What might be the best way to get to know your target users and stakeholders?
- What kind of social issues have you considered as part of your ideation process?
- What inspires your design solution?
- What are the limitations to your design?
- How might you communicate your strategy to your target users and stakeholders?

References

Koupf, D. (2017). Proliferating textual possibilities: Toward pedagogies of critical-creative tinkering. *Composition Forum*, *35*, 1–13.

Definition of a writing intensive course. (2010). Office of Undergraduate Education. University of Minnesota. Retrieved from http://archive.undergrad.umn.edu/cwb/definition.html

5

CULTIVATING RADICAL COLLABORATION IN TECHNICAL COMMUNICATION

Overview: At the core of design thinking, making, and social innovation is an indispensable spirit of collaboration. This chapter promotes the "radical collaboration" attribute in design thinking to empower multivocality and diverse contributions in user-centered design. Through a self-study of an interdisciplinary research collaboratory, this chapter reveals the challenges and strategies in cultivating collaboration in and outside of the academy. Following the case study, this chapter presents a set of scaffolding exercises for supporting collaborative projects in technical communication courses. Instructors and students can use these inclusive strategies to design collaboration teams and project workflow. For technical communication professionals, this chapter details the ways in which practitioners can harness the power that diverse perspectives bring to social innovation.

Collaboration in Technical Communication

Unlike most of my friends in the professional circles, my discovery of design thinking was not through software development or UX design contexts. I have instead learned about design thinking from my engagement with an interdisciplinary research group at the University of Minnesota. It was through this multi-year collaboration with researchers, designers, and administrators that I realized the true importance of design thinking. In the previous chapters, I have highlighted the ways in which design thinking as an innovation methodology can support active learning, social innovation, and community engagement. However, what has not been given enough attention in design thinking literature is its capacity for organizing teams and facilitating transformative collaborations. Hence, this chapter seeks to uncover how design thinking can afford human-centered collaboration to support social innovation efforts. Through a self-study of "radical collaboration," I highlight the facets of design thinking that mobilize teams for interdisciplinary research and design. First, I situate collaboration as an ongoing topic of interest in technical communication and show a trajectory in the field's discussions toward design-centric methodologies.

In reviewing our field's literature, Isabelle Thompson (2001) observed that "collaboration as a research issue and as practice seems firmly rooted in technical communication as a discipline (p. 167). For decades, collaboration has been formally taught and studied as a qualitative skill of technical communication. The "social turn" of the late 1970s brought about by writing and communication theorists was largely responsible for the increased investment in collaboration studies in technical communication. Writing studies scholars like Kenneth Bruffee (1984), Anne R. Gere (1987), and John Trimbur (1989) have influenced the early research directions for collaboration in academic and professional settings. Of note is Lisa Ede and Andrea Lunsford's definitive body of work (Ede & Lunsford, 1983, 1985; Lunsford & Ede, 1984, 1986), which was synthesized into their magnum opus, *Singular Texts/Plural Authors* (1990). Within the domain of technical communication, our field has seen attempts to capture the behavioral, social, and technological phenomena in collaborative activities, resulting in publications like the special issues of *Bulletin of the Association for Business Communication* (later *BPCQ*; Beard & Rymer, 1990), *Technical Communication* (Bosley & Morgan, 1991), and *Technical Communication Quarterly* (Burnett & Duin, 1993). Mary Lay (Schuster) and William M. Karis's (1991) edited collection, *Collaborative Writing in Industry*, provided additional perspectives and strategies learned from workplace collaborators.

As communication technologies evolve, technical communicators have focused on the influence of tools in collaboration and emerged as experts who facilitate team interactions and manage projects with emerging technologies such as content management systems, cloud repositories, open-source applications, and virtual team participation platforms. As evident in books like *Computers and Technical Communication* (Selber, 1997), *Technical Communication and the World Wide Web* (Lipson & Day, 2005) and *Digital Literacy for Technical Communication* (Spilka, 2010), research on collaboration and technical communication in the early 2000s was primarily driven by the affordances of the internet and the Web. Instructors were curious if and how digital technologies could better facilitate collaborations. For instance, Paul Benjamin Lowry, Aaron Curtis, and Michelle René Lowry (2004) studied emergent collaborative writing technologies and stressed that communication software serves as a mediator of successful collaborations. Organized panels and plenary sessions at various academic conventions like the annual meetings of Association for Teachers of Technical Writing (ATTW) and Council for Programs in Technical and Scientific Communication (CPTSC) frequently featured pedagogical innovations that leveraged the evolving functions of collaboration technologies.

Given the increased complexity that comes with technologized interactions in the work across branches of technical communication and adjacent fields, there are growing interests for new strategic methods for collaboration that can enhance product quality, improve team member experience, and reduce production cost. Approaches like lean and agile team management methodologies have been widely adopted as an integrated design and development process that yields success. In *Lean Technical Communication*, Meredith Johnson, Michele Simmons, and Patricia Sullivan (2018) presented case studies in technical communication program administration that demonstrated how lean methods can lead to sustainable programmatic design. In her business and technical communication courses, Rebecca Pope-Ruark (2012, 2014) used agile and scrum methodologies to help students communicate team goals and meet project requirements. It goes without saying, these approaches are becoming commonplace in technical communication programs and are certainly popular in industry practices.

UX consultant Jeff Gothelf (2017) took a close look at these methodologies and revealed their common underpinning values. Gothelf found lean and agile approaches to be informed by design thinking principles. After all, both lean and agile methodologies seek to promote a cyclical design process and support continuous improvement. These processes are actualized by solutions-driven design methods that prioritize rapid prototyping, testing, and iteration. Design thinking is a more pertinent framework for collaboration with its emphasis on human-centered design as a governing principle. It augments lean and agile methodologies with a focus on empathy and participatory design methods where the users' voice can be represented more prominently in the design outcomes.

Design thinking also confronts conventional ideologies in disciplinary collaboration. In "What Can Design Thinking Offer Writing Studies?", Jim Purdy (2014) contended that design thinking approaches challenge scholars to consider partnerships with those outside the "written word" tradition, and justify interdisciplinary collaboration. Interdisciplinary collaborations can open doors to new concepts, amended methods, and even new subfields that respond to modern needs. For technical communication, design thinking can take the already interdisciplinary profession to a new level of collaboration. In the following section, I discuss how design thinking provides a mechanism for productive collaboration while empowering collaborators to make meaningful contributions to their projects.

Design Thinking Attributes and Collaboration

The Hasso Plattner Institute for Design (n.d.) at Stanford University, commonly known as the d.school, has identified the following key attributes of design thinking to be applied in the problem-solving process:

- **Focus on human values:** Empathy for the people you are designing for and feedback from these users is fundamental to good design.
- **Show, not tell:** Communicate your vision in an impact-full and meaningful way by creating experiences, using illustrative visuals, and telling good stories.

- **Embrace experimentation:** Prototyping is not simply a way to validate your idea, it is an integral part of your innovation process. we built to think and learn.
- **Be mindful of process:** Know where you are in the design process, what methods to use in that stage, and what your goals are.
- **Bias toward action:** Design thinking is a misnomer, it is more about doing than thinking. Bias toward doing and making over thinking and meeting.
- **Craft clarity:** Produce a coherent vision out of messy problems. Frame it in a way to inspire others and to fuel ideation.
- **Radical collaboration:** Bring together innovators with varied background and viewpoints. Enable breakthrough insights and solutions to emerge from the diversity.

Stemmed from the phases of design thinking (empathize, define, ideate, prototype, test), these attributes can be used as guiding principles for collaboration in technical communication. These attributes promote human-centeredness and action-based problem solving; they inspire team members to experiment with ideas and cultivate creative solutions through unconventional processes. The radical collaboration attribute overturns traditional hierarchical structures in teams by fostering an equal sense of project ownership. This mindset defies typical top-down communication and workflow, and serves to empower marginalized voices in a team. Ideally, all team members should share similar decision-making power and stakes in the outcomes of the collaboration.

Indeed, the attributes of design thinking may boost the team collaboration experience. But since technical communication is (always) already a collaborative practice, how can design thinking enhance the collaborative process in technical communication contexts? Based on a collective investigation, Duin, Moses, McGrath, Tham, and Ernst (2017) have characterized the following tenets for radical collaboration: exposure, collaboration, invitation, suspension, sharing, and radical imagination. In essence, radical collaboration supports these tenets by:

- **Exposure:** Exposing participants to the complexities of problems regardless of experience.
- **Collaboration:** Resisting hierarchical structures; inviting and welcoming perspectives across disciplinary and institutional boundaries.
- **Invitation:** Inviting and welcoming perspectives that span theoretical, personal, and professional boundaries.
- **Suspension:** Suspending beliefs about knowledge boundaries; suspending judgment of people and ideas; suspending closure.
- **Sharing:** Exploring empathy together as a collaborative learning tool; sharing leadership, research, and teaching roles.

- **Radical imagination:** Inviting and activating radical change to what _____ (subject matter) can mean and be.

At its core, radical collaboration seeks to flatten power structures with the goal to harness collective creativity in addition to individual expertise. The outcomes of this design thinking-powered collaboration model include a diverse team with members who are comfortable offering differing perspectives and ideas, and an atmosphere that is conducive for imaginative solutions.

A Case of Radical Collaboration

> *Nota bene*: The following case study was performed in collaboration with my colleagues at University of Minnesota, and I especially want to acknowledge the intellectual contributions from the following scholars: Ann Hill Duin, Joseph Moses, Jeremy Rosselot-Merritt, Nathan Bolig, and Saveena (Chakrika) Veeramoothoo. The findings from this case study and the subsequent discussion were informed by the collective input by these scholars.

To examine the practical implications of radical collaboration, I showcase here a study of technical communication research collaboration where radical collaboration was employed. As I detail later in the methodology section, this study relied on an ethnographic method that leverages the nature of radical collaboration to perform critical self-evaluations. The collaborative team under study was the aforementioned research group that I was a part of at the University of Minnesota—the (then) Wearables Research Collaboratory.

Officially launched in the Fall of 2015, the Wearables Research Collaboratory (WRC) was first piloted as a special interest research group by two professors and two graduate students—I was one of them—in the Department of Writing Studies in 2014. The group started out with a focused study of the wearable computer Google Glass XE edition, then quickly evolved into an open collaboration unit where anyone from across the department and university could join to participate in research. The exigence came from our learning of the growing interests in wearable technology across many disciplines, and the guiding principles of design thinking, which we were just beginning to grasp, that encourage multidisciplinary perspectives to inquiry. On the 2015 website (Figure 5.1), the WRC declared that it

> is an open collaboration and research space for wearables-related initiatives, projects, and ideas stemming from the burgeoning interest in wearables and their impact on users and their work. As a collaboratory, we represent an incubator for bold ideas, an environment where participants explore emerging wearables and share empirical direction for investigating the challenges and opportunities these technologies represent.

FIGURE 5.1   A screenshot of the WRC website in 2015

Source: Archived by author

Projects led by graduate students and faculty alike addressed:

- Perspective shifting and user experience
- New audience and digital rhetoric theory
- Cultural perceptions of wearable technologies
- The rhetorical situation through virtual and augmented realities

The notion of a "collaboratory" alludes to the combination of collaboration and laboratory, a space for experimentation and learning. Through collaborative problem-scoping, idea generation, and solution finding and innovation, WRC members focused on understanding the value of wearable and emerging technologies, and sharing empirical direction for investigating the challenges and opportunities these technologies present. Methods for engagement included sharing an open meeting agenda where all members could add or edit agenda items; inviting speakers and hosting virtual meetings with subject experts outside the collaboratory; giving presentations at departmental, university, as well as national and international conferences; hosting "pop-up" events across the university; touring research centers; mentoring undergraduate researchers; and curating teaching resources and tutorials online.

Radical collaboration was enacted through the shared leadership in team organizing (e.g., meetings, decision making). Regular meetings were chaired by different members of the collaboratory each week, with ad-hoc groups, formed out of shared interests, that held self-organized meetups. When workshopping ideas, members abided by the radical collaboration tenets and suspended judgement and closure; this led to a more genuine sharing of ideas with minimal fear of premature critique. When sharing research results and publication opportunities, the collaboratory members upheld an always-already collaborative belief so authorship was negotiated based on commitment rather than power relations.

Collaborative Autoethnographic Methodology

A study on collaboration calls for a collaborative research method. Unlike the interviewing method featured in the previous chapters, this study employed an emerging collaborative investigation methodology that involved both individual and collective reflections, namely collaborative autoethnography (CAE). As a methodology, CAE is an ensemble approach to individual reflections that invites collective sharing and probing, meaning-making, and composition. This methodology aligns with the guiding principles of design thinking and, more importantly, radical collaboration. According to Heewon Chang, Faith Ngunjiri, and Kathy-Ann Hernandez (2013), CAE "focuses on self-interrogation but does so collectively and cooperatively within a team of researchers" (p. 21). CAE is "a qualitative research method in which researchers work in community to collect their autobiographical materials and to analyze and interpret their data collectively to gain a meaningful understanding of sociocultural phenomena reflected in their autobiographical data" (pp. 23–24).

What makes CAE unique to technical communication research is that it diversifies the researchers' and their participants' viewpoints. As Chang et al. (2013) stated, "the combination of multiple voices to interrogate a social phenomenon creates a unique synergy and harmony that autoethnographers cannot attain in isolation" (p. 24). WRC members have worked as a collaborative to articulate research questions of relevance both to individual and collective scholarship. To exemplify the notion of radical collaboration, my collaborators and I—during the course of this self-study—have worked to shift writing, reflections, and observations from individual to collective interpretation, so we can achieve what CAE researcher Judith Lapadat (2017) called, "a shift from individual to collective agency" (p. 1).

CAE supports an iterative data collection and interpretation process. During the process, several methods may be used to facilitate individual and collective reflections. In studying the WRC, we employed a survey questionnaire (see Appendix A) for individual reflections and a focus group-style discussion for collective reflections. To capture the overall experience of WRC members as having undergone the radical collaboration process, this self-study was conducted at the end of 2017 spring term (end of academic year). Figure 5.2 sums up the data collection and analysis process in this study.

The survey questionnaire was designed to gather qualitative data about the WRC members' experience in the spring semester rather than to generate generalizable findings about radical collaboration. In the questionnaire, WRC members were asked to describe their experience with the collaboratory, and the ways in which the six tenets of radical collaboration had been actualized. Besides providing written responses, WRC members were also asked to score the degree of actualization for each of the six dimensions (from the scale of 0–5, with 0 being not actualized and 5 being fully actualized). The questionnaire was given to all WRC members prior to the last WRC meeting of the term. After completed the questionnaire individually, the participants discuss their responses to each question during the focus group meeting.

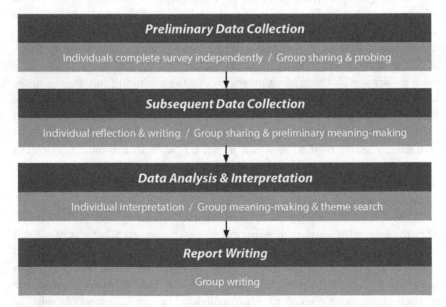

FIGURE 5.2 The iterative process of collaborative autoethnography in this study

Results

A total of 10 WRC members (out of 11 active members) participated in this self-study. Table 5.1 shows the academic "rank" of the participants. Participants included three undergraduate research assistants who had worked on various projects with WRC graduate students and faculty members during the spring 2017 semester. Participants were asked to focus their responses on their experience in the spring semester so they all could recognize specific references or ascribed events.

The open-ended questions on the survey questionnaire allowed participants to contemplate on their respective experiences in radical collaboration, with prompts to probe further elaboration. Again, CAE supports individual reflections and collective meaning-making. Participants used the survey questions as conversation starters during the focus group meeting. The first question sought to capture the participants' overall experience. Table 5.2 shows the participants' description in a word or a phrase.

Aside from Ann's response—which provided a metaphor for collaborative work—and Laura's one-phrase summary of her major project with the collaboratory, all other responses focused on energy and engagement. Participants mainly reported that the WRC was a place where they felt energized, welcomed, and engaged.

The remaining questions asked participants to provide detailed responses on how they felt each of the six tenets of radical collaboration was achieved or not during the spring semester. In the following subsections, I provide the results from the group sharing and collective meaning-making. Each subsection is a tenet of radical collaboration with representative quotes by the participants to exemplify the workings of the tenet.

1. Invitation

The tenet of invitation seeks to welcome perspectives that span theoretical, personal, and professional boundaries. For undergraduate research assistants (RAs),

TABLE 5.1 Participants' academic rank at the time of survey

Name	Academic rank	# of semesters in the WRC
Ann	Professor	5
Joe	Senior lecturer	5
Megan	PhD candidate, 4th year	5
Jason	PhD candidate, 3rd year	4
Nathan	PhD student, 1st year	1
Jeremy	PhD student, 1st year	1
Saveena	PhD student, 1st year	1
Bilal	Undergraduate researcher	1
Laura	Undergraduate researcher	1
Alexander	Undergraduate researcher	2

TABLE 5.2 Participants' responses to Question 1

Please use a word or a phrase to describe your overall experience as a member of the WRC this semester.

Ann	"Consider a hammock. With design thinking, we are suspended by tension, often in directions with contrasting needs (faculty/student; grad/undergrad; writing/TPC; independent/collaborative). Yet such tension is required to exist, to move, to suspend. . . . Like a hammock, one has to 'plunge' a bit into the WRC, and immediately you find yourself moving with it."
Joe	"Emergent creativity—as satisfying as sitting down to write and wondering what's going to happen."
Megan	"Energizing."
Jason	"Productive. I feel that I have done/written a lot as a member here. WRC truly is an incubator for ideas and research activities that span beyond my regular graduate curriculum."
Nathan	"Equitable and enthusiastic."
Jeremy	"Collegial."
Chakrika	"Collaborative work."
Bilal	"Welcoming of all ideas, no matter how abstract."
Laura	"Virtual reality exploration."
Alexander	"Engaging!"

invitation was perceived as an effort that encouraged them to participate in graduate-level research activities, including inquiries that were deemed ambiguous at first. In his response, Bilal (RA) wrote:

> I feel like the RAs were included in a lot of the graduate activity and discussion which ushered a sense of interest for us.

For the graduate-student participants, invitation was perceived as implicit and a welcoming effort. They reported that even new members of the collaboratory have equal opportunities to contribute to the overall mission of the organization. For faculty members, to achieve invitation means to actively keep the doors of the collaboratory open, and seek out opportunities where invitations can be extended. It also means to resist conventional power relationships in formal hierarchies (faculty-student, graduate-undergraduate, older-younger scholar, etc.) and allow everyone in the collaboratory a fair opportunity to participate in the ongoing discussions and decision-making processes.

2. Sharing

Sharing in radical collaboration means nurturing empathy to foster authentic interactions. When ideas are shared and workshopped, collaboratory members sought to empathize with the sharer's intentions. From the participants' reflection, sharing was deemed an important tool to flatten power hierarchy by giving all members the opportunity to share leadership. Each week during the semester, WRC members took turns to chair the collaboratory meeting while everyone collectively contributed to generating and documenting the meeting agendas and records. Sharing was also achieved by the means of collaborative technology, as graduate student Megan wrote:

> We are very sharing-heavy, in general, in terms of how frequently we email each other, how we share items on the agenda each week, and how we use Google Drive. I really appreciate how, even with all of this sharing, nobody seems to feel proprietary, which shows me just how important a sense of trust is in a group like this one.

For the undergraduate RAs, sharing was a way for them to learn by doing. For graduate students, sharing was about opening up spaces for critique and taking a leading role at times so others may follow successful examples. For faculty members, sharing afforded more scholarly activities that were difficult to spearhead on their own; examples included the "pop-up" events the WRC hosted on campus and other graduate student co-teaching activities.

3. *Collaboration*

Certainly, for radical collaboration, the tenet of collaboration was most prominent. As presented in Table 5.3, the participants rated collaboration with the highest median and mean scores. Genuine collaboration resists hierarchical structures, invite and welcome perspectives across disciplinary and institutional boundaries. In their responses, the undergraduate RAs noted how collaboration was a constant effort they saw in the collaboratory, as they were paired with graduate students and faculty members to assist in respective teaching and/or research activities. Graduate students and faculty members found that collaboration required a certain level of energy to actively engage with one another, while keeping individual research interests. This almost became challenging for senior lecturer Joe, who wrote:

> There were so many opportunities to collaborate that I found myself withdrawing a bit so as not to overcommit.

At the same time, however, collaboration was what kept the collaboratory's momentum throughout the semester and helped its members stay productive. Undergraduate RA Laura stated:

> Collaboration for me was also achieved through projects that I did for my classes that I was able to leverage/relate to my work with the Collaboratory. . . such as a brochure I made [for a class], and interviewing people for the podcast that I'm currently working on.

Laura's testimony was also an instance of how the work at the collaboratory added value to her classes. In fact, her collaboration with WRC member Megan has led to a podcast project that Laura created for a class in science and technology.

4. Radical Imagination

Radical imagination means inviting and activating radical change to what any subject matter can mean and be. In this case of WRC, we radically imagined what technology-enhanced teaching and learning can mean and be in higher education. In this self-study survey, radical imagination received the lowest median score among all tenets of radical collaboration. To most WRC members, what it means to be "radical" was still subjective to the work of each individual. The undergraduate RAs noted that it was difficult to recognize radical imagination as they were unfamiliar with many concepts used in the works of graduate students and faculty members. Graduate students, however, saw the works of undergraduate RAs as radical—such as their pop-up event and successful acquisitions of seed grants by the college. Graduate student Nate noted:

> There are some incredible new directions that the RAs have brought into play and I am seriously impressed with their enthusiasm and work ethic for making proper use of the funding and technology we are so lucky to use! The RAs are the kindling to the fire of the group in my opinion.

Faculty members saw radical imagination as both an exploration as well as a practical organizational challenge. Ann noted that collaborative leadership outside the collaboratory are necessary catalysts for radical imagination.

5. Suspension

Suspension requires a focus on problem development. It was also one of the most difficult tenets to be actualized, according to the ratings provided by WRC members. Suspension had received the lowest mean score among all tenets of radical collaboration. Most members thought that the collaboratory as a whole had done well in critiquing each other's work while being supportive. In my own reflection, I wrote:

> We do pretty well in supporting and critiquing one another's ideas. Although, many times, I do see us going down a groupthink spiral. . . but we recover quickly when it comes to the practical issues with our ideas. We are slow to judge, but considerably quick to agree. The beautiful thing, however, is that we offer ourselves to participate in things we agree on.

Faculty member Ann had expressed a similar sentiment, noting:

> We locked in to studying VR quite early in the process rather than suspend-
> ing this direction or at least entertaining many other directions. We need
> to keep focused on 'problem development' throughout our collaboration.

Overall, suspension was an underlying warrant of the collaborative work in the
collaboratory that was hard to actualize. However, when complemented by other
tenets of radical collaboration, such as sharing and collaboration, suspension contributed to the momentum of the collaboratory. It created a culture of acceptance
and kindness toward others' scholarship that is arguably rare in academia.

6. Exposure

Exposure is a tenet that treats all collaborators as equally worthy of participating in
problem scoping. It means exposing participants to the complexities of problems
regardless of experience. WRC members all agreed that they were exposed to
something they have not encountered before participating in the collaboratory
in spring semester. The RAs noted their eyes were opened to various aspects of
technical and professional communication through the work of the collaboratory.
Alexander, a second-year RA, reported:

> Exposure was extremely important for what I have learned over the last
> semester. My first semester here, I was able to experience many technical
> writing ideas and I learned a few things for our collaboration. But with
> this new semester, I was able to connect the ideas from this semester to my
> major more, and provide much greater context to what I learned here in
> the collaboratory.

Graduate students perceived exposure as experience cross/interdisciplinary scholarship. Graduate student Chakrika noted:

> I think our visiting people doing similar work as ours counts as exposure to
> build our research network. I think that we also achieved exposure to new
> technologies and tools for potential future research or teaching use.

Faculty members, too, saw connecting with units outside of the collaboratory
and department home as a way of exposing student members to common issues
(mostly technology-related), breaking down academic silos, and capitalizing on
talents outside the immediate reach of the field.

To tabulate the degree to which each radical collaboration tenet is actualized
in the members' experience, each member was asked to rank their individual

experience of the dimensions on a scale of 0 to 5 (lowest to highest). Table 5.3 and Figure 5.3 show the median and mean scores for each dimension.

Collaboration, sharing, and invitation were ranked as the top three dimensions, followed by radical imagination, exposure, and suspension accordingly. The lowest mean score was 3.4 (suspension) and its median was 4.0. On a 5-point scale, the scores reflect a high degree of actualization across the six dimensions. Their experience in the WRC was thus representative of radical collaboration. Certainly, the six radical collaboration tenets do not need to be equally actualized in order to achieve optimal collaborative experience; yet, based on the survey feedback, they have to be present to manifest radical collaboration in the collaboratory.

While this CAE self-study and narratives have been limited to the context of academic research in technical communication, I believe radical collaboration be

TABLE 5.3 Degree of actualization of radical collaboration dimensions

Dimension	Mean	Median
Collaboration	4.4	4.5
Sharing	4.3	4.5
Invitation	4.1	4.0
Radical imagination	3.7	3.5
Exposure	3.6	4.0
Suspension	3.4	4.0

Range = 0 to 5 (0 being lowest; 5 being highest)

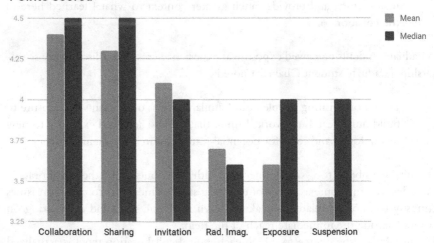

FIGURE 5.3 Comparison of degree of actualization (mean and median) across six radical collaboration dimensions

useful for industry application. In the remaining of this chapter, I advocate for the adoption of this practice by academic programs and practitioners alike who wish to build and support a culture of radical collaboration. The radical collaboratory model (see Figure 5.4) can be deployed by considering three main conditions in the academy: disciplinary, institutional, and programmatic.

Supporting a Culture of Radical Collaboration

Figure 5.4 serves as a guiding visual when planning and creating radical collaboration within a field, an institution, or a graduate program. I offer some action items under each juxtaposition of condition and radical collaboration component. These suggestions are outlined in Table 5.4.

At the disciplinary level, our field needs to continue to strengthen its initiatives in encouraging cross-disciplinary scholarship and creative activities. While national meetings and international academic conventions are common ways to welcome scholars to participate in knowledge creation and dissemination,

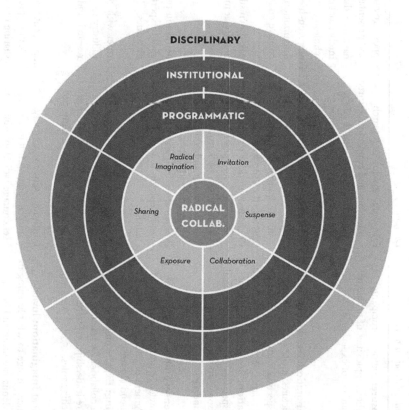

FIGURE 5.4   A model for radical collaboratory across programs, institutions, and disciplines

TABLE 5.4 Strategies for cultivating radical collaboration across disciplinary, institutional, and programmatic contexts

Radical collaboration tenets	Disciplinary	Institutional	Programmatic
Exposure: Expose participants to the complexities of problems regardless of experience.	Encourage sharing of research questions and continuous efforts pertaining to the vitality of the field regardless of researcher/scholar's experience.	Support cross-unit scholarship and presentations of complex narratives that affect the intellectual community as a whole.	Develop initiatives to expose new and current members to shared interests and wicked problems with ambiguous solutions.
Collaboration: Resist hierarchical structures; invite and welcome perspectives across institutional boundaries. Welcome co-authorship.	Welcome grassroot efforts to shed light on new perspectives and initiate appropriate changes to outdated methods and operations.	Foster non-hierarchical organization to research and creative endeavors.	Create collaboratories ungoverned by existing power structures to investigate complex issues.
Invitation: Invite and welcome perspectives that span theoretical, personal, and professional boundaries.	Keep an open door to scholars, researchers, and practitioners from various backgrounds to participate in and contribute to disciplinary activities.	Break down silos; promote shared inquiry at multiple structural levels.	Inspire mentor–mentee and peer collaboration; disrupt conventional scholarly expectation.
Suspense: Suspend beliefs about knowledge boundaries; suspend judgment of people and ideas; suspend closure.	Problematize disciplinary norms and suspicion toward radical ideas.	Question institutional ideologies about expertise, skills, and identities.	Allow learners to experiment and pursue new endeavors; create platforms for sharing and workshopping ideas.
Sharing: Explore empathy together as a collaborative learning tool; share leadership, research, and teaching roles.	Devise mechanisms for vertical communication (students to scholars to field leaders); create platforms for open forums and discussions.	Partner with community and industry to share leadership in research and teaching.	Emphasize empathy in collaborative learning; hold conversations about failures and successes.
Radical Imagination: Invite radical change to what learning in academia can mean and be.	Cultivate a sense of curiosity and continuous improvement; encourage radical imagination.	Implement frameworks for defining, ideating, prototyping, testing, and sharing of radical ideas.	Incite bold yet ethical methods and experiments for transforming what it means to practice, teach, and learn.

field leaders should pay attention to emerging needs for radical collaboration. Efforts such as Digital Rhetoric Collaborative, HASTAC, and the journals *Digital Humanities Now* and *Hybrid Pedagogy* are exemplary cross-disciplinary initiatives that foster intentional collaborative scholarship and widen our field's approach to research, teaching, and learning.

At every home institution, I recommend installing infrastructures that support cross-unit, non-hierarchical work around complex challenges that define and redefine the purpose of the institution. At the University of Minnesota, for example, efforts such as the Grand Challenges curriculum, Interdisciplinary Collaborative seed funding, and the Institute for Advanced Study can break down silos and press for new ideas and challenge the institutional ideologies about expertise and identities. I particularly admire the Grand Challenges initiative as a way to promote radical collaborative scholarship not just among faculty members but with students to create deep engagement with real-world challenges.

Finally, at the program level, there needs to be mechanisms that help prepare students to become radical collaborators at the institutional and disciplinary levels. In our collaboratory, we have visited with mechanical engineering professors working on developing "smart" fabrics for medical and rehabilitative purposes; met with researchers at the Wearable Technology Lab where they are developing new on-body technologies for sensing, actuation, and communication; met with graduate students at the Medical Device Center to explore the use of virtual reality for understanding how medical devices will later work during specific surgeries; hosted virtual conference sessions with companies developing smart glasses for use in field contexts and with a technical communication professor studying diabetes wearables; and interfaced with members of the University of Minnesota's Technical Communication Advisory Board comprised of professionals in a number of different industries. These experiences—in addition to the research that the graduate students conducted for their own projects with the help of the undergraduate RAs—were particularly relevant to workplace collaboration because they allowed us to see how problems are being solved through interdisciplinary collaboration; and, on many levels, these provide the opportunity to be a part of purposive collaboration ourselves. That purposive collaboration, in turn, benefits our pedagogical perspective when we teach courses in technical and professional writing for students who will work in any number of professions in their futures: technical communication, engineering, computer science, dietetics and food science, law, economics, and a host of others.

Certainly, the WRC is just an instance of such effort. I recommend that local programs create spaces and collaboratories for students to collaborate with each other and with faculty members; experiment, share, and workshop new ideas; and practice empathy. Student-led publications and research activities, peer reviews, and parlor-style presentations are typical examples of such effort. Pushing even further, I implore local programs to consider collaborative scholarship—like co-dissertations—as a way to embrace radical collaboration.

Designing Collaborative Project Workflow With Radical Collaboration Tenets

The tenets of radical collaboration support not just large-scale collaborative efforts but also smaller teams and partnerships in designing project workflows. As April Greenwood, Benjamin Lauren, Jessica Knott, and Danielle Nicole DeVoss (2019) found in their collaborative study, "Effective facilitation that incorporates building empathy between team members is one way that design thinking can help groups of diverse people be creative and innovate together" (p. 414). Using the guiding questions below, team members may identify their shared values, understand the feasibility of their collaboration, and co-create guiding principles that serve the needs of the collaboration.

Pre-Audit Reflection

Before attempting the radical collaboration audit, have an open discussion with your collaborator(s) about your expectations using the following prompts:

- What is the purpose of this collaboration?
- What is the scope of work and associated time frame for this collaboration?
- How should personal interests and goals for this collaboration be accounted?
- What are the expected outcomes of this collaboration?

Radical Collaboration Audit

Upon reflecting on the logistical aspects of your collaboration, use the following list of questions to audit your partnership or team's readiness to engage radical collaboration. These questions are meant to prompt discussions rather than solutions.

Post-Audit Discussion

After completing the auditing worksheet, collaborators may design an agreeable workflow for radical collaboration. Use the following prompts as guidelines:

- **Questions → ideas:** How do questions get taken and responded to? When ideas are forming, where do they get scribed or documented? How might collaborators access ideas in formation?
- **Ideas → discussions:** How do ideas get included on the discussion agenda? How are ideas grouped or separated? How do these affinities/differences get evaluated?
- **Discussions → plans:** How do discussion items become actionable plans? How are plans named or labeled? How will progress from each plan be reported?

TABLE 5.5 Radical collaboration auditing worksheet

Tenet	Question
Exposure	1. Is exposure to a greater level of complexity (for problem definition, skill development, etc.) necessary for this collaboration?
Exposure	2. How might collaborators communicate their comfort level with exposure to new complexity or unknown?
Collaboration	3. How should the team be maintained?
Collaboration	4. What roles are necessary for team maintenance? (i.e., scheduler, documentarian, or other shared roles.)
Invitation	5. How might new members participate in the ongoing collaboration?
Invitation	6. How might new perspectives be invited?
Suspense	7. How should collaborators practice active listening?
Suspense	8. How are ideas assessed and critiqued?
Sharing	9. How can different roles (see Question 4) be rotated or shared?
Sharing	10. How is ownership of the collaborative project defined?
Radical Imagination	11. How should ideas take form?
Radical Imagination	12. How might failures be embraced?

- **Plans → prototype(s):** Who will decide when plans are ready to be prototyped? What resources would be necessary for building these prototypes? Who will procure these resources?
- **Prototype(s) → testing:** Where will prototypes be stored and who has access to them? How will testing be conducted? How much testing would be needed?
- **Testing → implementation:** How will the results of testing be interpreted? How will the test results be used in iterating the prototype? Who shall decide when the revised prototype is ready for implementation? Who will bear the cost for implementation? How will successes and failures be embraced?

Summary and Takeaways

Collaboration is innate to technical communication. Contemporary problems call for innovation in collaborative practices. This chapter recommended a "radical collaboration" approach afforded by design thinking as a way to decentralize power relations in interdisciplinary collaboration. Using collaborative autoethnography methods, I presented a self-study of an academic research collaboratory and revealed how tenets of radical collaboration can be actualized. The chapter concluded with a three-tier framework for supporting a culture of radical collaboration in the academy that can also be adopted by industry programs. Key takeaways from this chapter are:

- Collaboration needs to be continually examined in technical communication.
- Design thinking can enrich collaboration by fostering a "radical" approach to team organizing.
- Collaborative autoethnography can be an effective method for team's self-examination.
- Radical collaboration can be manifested across disciplines, institutions, and programs through the tenets of exposure, collaboration, invitation, suspension, sharing, and radical imagination.

Learning Activity: Considering Dilemmas in Radical Collaboration

The tenets of radical collaboration emphasize openness and shared ownership of ideas and resources. Now, what happens when there are conflicts of interest between collaborators? Consider the following situation:

> Mandy, a graduate student who has just completed her coursework is heading into her "dissertation phase" and is writing a National Science Foundation grant application to support her summer research. Given the focus of her dissertation topic, which is an investigation of wearable technologies in the workplace, Mandy is excited to share her grant application with the members of her research collaboratory so they may provide feedback.
>
> During the research collaboratory's meeting, Mandy is surprised to learn that two other grad students in the collaboratory, Jasper and Sheila, are working on applying for the same grant. Knowing that the grant sponsor does not usually fund multiple projects from the same institution, Jasper and Sheila ask Mandy if she would like to join their grant proposal as a third author. Mandy is torn between her preference to receive the grant on her own (which would mean larger funds) and her fear of losing the grant to Jasper and Sheila if she went on to submit the application separately.

Discussion questions:

- What would you do if you were Mandy? Why?
- What are some helpful resources or strategies to resolve a situation like this?
- Was competition inevitable? Is this competition healthy for the collaboratory? Why?
- What contextual factors would you consider from this scenario? How do the tenets of radical collaboration support (or not) the decision-making process?

References

Beard, J. D., & Rymer, J. (1990). The contexts of collaborative writing. *Business and Professional Communication Quarterly, 53*(2), 1–3.

Bosley, D. S., & Morgan, M. (1991). Special issue: Collaborative writing: Introduction. *Technical Communication, 38*(4), 474–475.

Bruffee, K. (1984). Collaborative learning and the "conversation of mankind." *College English, 46*(7), 635–652.

Burnett, R. E., & Duin, A. H. (1993). Collaboration in technical communication: A research continuum. *Technical Communication Quarterly, 2*(1), 4–20.

Chang, H., Ngunjiri, F., & Hernandez, K-A. C. (2013). *Collaborative autoethnography.* Walnut Creek, CA: Left Coast Press.

Duin, A. H., Moses, J., McGrath, M., Tham, J., & Ernst, N. (2017). Design thinking methodology: A case study of "radical collaboration" in the wearables research collaboratory. *Connexions: International Professional Communication Journal, 5*(1), 45–74. Retrieved from https://connexionsjournal.org/wp-content/uploads/2019/12/duin-etal.pdf

Ede, L., & Lunsford, A. (1983). Why write. . . together? *Rhetoric Review, 1*(2), 150–157.

Ede, L., & Lunsford, A. (1985). Let them write—together. *English Quarterly, 18*(4), 119–127.

Ede, L., & Lunsford, A. (1990). *Singular texts/plural authors: Perspectives on collaborative writing.* Carbondale, IL: Southern Illinois University Press.

Gere, A. R. (1987). *Writing groups: History, theory, and implications.* Carbondale, IL: Southern Illinois University Press.

Gothelf, J. (2017). *Lean vs. agile vs. design thinking: What you really need to know to build high-performing digital product teams.* New York, NY: Sense and Respond Press.

Greenwood, A., Lauren, B., Knott, J., & DeVoss, D. N. (2019). Dissensus, resistance, and ideology: Design thinking as a rhetorical methodology. *Journal of Business and Technical Communication, 33*(4), 400–424.

Hasso Plattner Institute for Design. (n.d.). Design thinking bootleg. Retrieved from https://dschool.stanford.edu/resources/design-thinking-bootleg

Johnson, M., Simmons, M., & Sullivan, P. (2018). *Lean technical communication: Toward sustainable program innovation.* New York, NY: Routledge.

Lapadat, J. C. (2017). Ethics in autoethnography and collaborative autoethnography. *Qualitative Inquiry, 23*(8), 589–603.

Lay, M. M., & Karis, W. M. (Eds.). (1991). *Collaborative writing in industry: Investigations in theory and practice.* Amityville, NY: Baywood Publishing.

Lipson, C., & Day, M. (Eds.). (2005). *Technical communication and the World Wide Web.* Mahwah, NJ: Lawrence Erlbaum Associates.

Lowry, B., Curtis, A., & Lowry, M. R. (2004). Building a taxonomy and nomenclature of collaborative writing to improve interdisciplinary research and practice. *International Journal of Business Communication, 41*(1), 66–99.

Lunsford, A., & Ede, L. (1984). Audience addressed/audience invoked: The role of audience in composition theory and pedagogy. *College Composition and Communication, 35*(2), 155–171.

Lunsford, A., & Ede, L. (1986). Why write. . . together: A research update. *Rhetoric Review, 5*(1), 71–81.

Pope-Ruark, R. (2012). We scrum every day: Using scrum project management framework for group projects. *Journal of College Teaching, 60*(4), 164–169.

Pope-Ruark, R. (2014). Introducing agile project management strategies in technical and professional communication courses. *Journal of Business and Technical Communication, 29*(1), 112–133.

Purdy, J. (2014). What can design thinking offer writing studies? *College Composition and Communication, 65*(4), 612–641.

Selber, S. (1997). *Computers and technical communication: Pedagogical and programmatic perspectives*. Greenwich, CT: Ablex/Greenwood Publishing Group.

Spilka, R. (Ed.). (2010). *Digital literacy for technical communication: 21st Century theory and practice*. New York, NY: Routledge.

Thompson, I. (2001). Collaboration in technical communication: A qualitative content analysis of journals, 1990–1999. *IEEE Transactions on Professional Communication, 44*(3), 161–173.

Trimbur, J. (1989). Consensus and difference in collaborative learning. *College English, 51*(6), 602–616.

CONCLUSION

Disrupting and Innovating in Technical Communication Through Making and Design Thinking

Overview: This book closes with an invitation to students and industry professional readers to create waves and cultivate change in technical communication by means of design thinking and making. It motivates them to explore the human-centered problem-solving orientation design thinking affords for technical communication practices. It challenges them to cause productive disruption, shaking up the ways we see and do technical communication, and generates new possibilities in user-centered design.

Think. Do. Make a Difference

My parents are old-school pragmatists—they have taught me to always recognize my strengths and do what is right and reasonable to achieve my goals. When I told them I wanted to teach in higher education, their response was rather succinct: find your purpose, know your limits, make a difference. Honestly, I thought the advice was cliché but not impassable.

I relived my parents' mantra when I was completing my master's degree at St. Cloud State University (Minnesota); my alma mater launched an expansive campaign to reposition itself as an active-learning university. The campaign slogan—"Think. Do. Make a Difference"—was designed to signal the hands-on, problem-based learning that students may experience regardless of their chosen area of study. To me, the slogan served more than a catchphrase to attract donors; it had prompted me to look into my teaching practices and reflect how I am creating opportunities for real impact in my students. In what ways can I better equip students to be change agents in a highly technological yet socially tensed society? This motivation for pedagogical research was renewed when I found myself listening with awe to Joyce Carter's address to about 3000 attendees at the annual

Conference on College Composition and Communication in 2016. Similar to the morale invoked by SCSU's campaign slogan, Carter called scholars to assume active roles in creating waves and cultivating change in the academy and beyond.

The discussions and studies in this book sought to underscore the values of making and design thinking in enabling positive change. This change can start from our classrooms. Whether it is a first-year survey course or a graduate pro-seminar, students can be afforded the knowledge and tools to engage with issues they consider important. Whether we all have access to fancy design technologies in makerspaces or simply low-fidelity materials for prototyping, we can still practice user-centered, mindful design via the design thinking mindset and methodology. And as our field scholarship shows, a productive way to achieving meaningful learning is through collaboration. Design thinking fosters *radical* collaboration and levels up how teams and organizations across different tiers interact. Echoing Carter's call for disruptive innovation, this closing segment situates design thinking and making as productive approaches to innovation in technical communication.

Design Thinking as Productive Disruption

As I have worked to demonstrate in this book, design thinking serves a crucial role in human-centered technology development. This is particularly important for designed solutions that may be perceived as radical or disruptive breakthroughs. The term "disruption" in business and entrepreneurial settings can be traced to Joseph Bower and Clayton Christensen's (1995) theory of disruptive innovation. Here I borrow Bower and Christensen's term to consider the role of design thinking in facilitating innovations that defy conventions and advocate for users.

But first, what counts as a disruptive innovation? According to the Christensen Institute (www.christenseninstitute.org/), disruptive innovations are not upgraded technologies that make good products better. Disruptive innovation is about making innovative products and services more affordable and accessible to a larger population. TurboTax is a featured example by the Christensen Institute (Fogg, 2018). By not just lowering the cost of tax preparation (thus more affordable) but also making the preparation process more user-friendly (accessible), TurboTax disrupted the accounting services industry and innovated a new way for individuals to prepare their own tax returns. As opposed to "sustaining technologies," which aim to reinforce market status quo, disruptive technologies "fundamentally change the way their users accomplish a goal, just as mobile phones have changed the way people use computing functionality" (Fogg, 2018, n.p.).

Yet, many tech companies attempting to update consumer technologies or reinvent customer experience are trapped by the conventional thinking that focused mainly on the technology itself. As the Google Glass example I featured in Chapter 3 shows us, technological innovations fall short when they lose sight of the users and their user experience. Bower and Christensen (1995) argued that in order to define the significance of disruptive technology, creators need "to ask the right people the right questions about the strategic importance of the disruptive technology. Disruptive technologies tend to stall early in strategic reviews because managers either ask the wrong questions or ask the wrong people the right questions" (n.p.). The "wrong people" include early adopters and designers who are passionate about emerging innovations. According to Bower and Christensen, these people do not provide fair feedback on the disruptive innovation.

Design thinking can be the salvation to disruptive innovators. Since design thinking advocates for well-defined problems and empathy toward users as the starting points for innovation, it does not seek to create disruptive solutions for disruption's sake. The user-centered methods supporting the design thinking process can help entrepreneurs and designers keep focus on the users and related stakeholders, not just those who are immediately available or lighthouse consumers. Furthermore, the iterative nature of design thinking ensures that designers do not get taken away by technological features but instead grounding their solutions in user requirements and feedback. At the same time, the collaborative and radical imagination attributes in the design thinking mindset motivate designers to exercise creativity mindfully—without losing sight of the actual user. In other words, design thinking stimulates *productive* disruption from the outset. Productivity here does not mean generativeness but effective, creative, and valuable innovation.

For technical communicators, the productive disruption bolstered by design thinking is key to answering Carter's appeal. Whether we are scholars, instructors, practitioners, or somewhere in between, we are all challenged by modern problems to innovate new approaches to communicate, to teach and learn, to research, and to organize ourselves. As we have seen through recent events, technical communication problems span beyond the *technical* aspects of communication. Facts and data are manipulated as "fake" information to cause conflict and social discord. Communication technologies are used as tools to segregate communities. Technical communicators are hence called to use our knowledge and skill sets to advocate for positive change. We are to "shift out of the neutral" (Shelton, 2019) to center social justice advocacy. As Carter put it,

> Advocacy means looking at the landscape, identifying the influencers and decision makers in the realm of culture, economics, and politics, and using our skills of persuasion, communication, and research to compel them to give us support, money, laws, and cultural capital.
>
> *(2016, p. 388)*

Carter regarded advocacy as one of the two modalities for socially responsive engagement, the other is making.

Making as a Disruptive Culture

As design thinking facilitates disruption, making sparks revolution. As I have briefly covered in Chapter 2, making is born of a resistance culture, and anti-technocentrism fueled the DIY movement. However, the 1980s global economic crisis was most instrumental in turning making from a hobby into a necessity. The idea of a dedicated (maker)space for informal personal learning and uncovering shared resources was born of this condition. Will Holman (2015), a general manager of a Baltimore-based makerspace, has found some early makerspaces formed by the public sector:

> In London, during the deep recession of the early 80s, voters elected a leftist city council to protest the austerity policies of the Thatcher government. Labour politicians quickly set up the Greater London Enterprise Board, which in turn established five Technology Networks with a budget of £4 million. These facilities, direct antecedents to modern makerspaces, were shared machine shops that aimed to democratize the means of production and access to education for unemployed manufacturing workers.
>
> *(Holman, 2015, n.p.)*

These members of a community of making shared an understanding of mutual ownership. Citing technology scholar Adrian Smith (2014); Holman (2015) reported that participants developed various prototypes and initiatives that created the idea of an "open access product bank" that distributed profits from inventions directly back to the members of the makerspace. Fast forward 20 some years, the modern maker culture has grown out of the mechanical industry into the tech industry while keeping the early synergy. Social media giant Facebook, for instance, integrates intentional makerspaces (called hackerspaces) to spark innovative ideas. First took place in 2007, the annual Facebook "hackathon" is designed for software engineers to move out of their regular silos or workgroups to collaborate with those they do not typically work with for the purpose of radical innovation. On a company page, former Facebook engineering director Pedram Keyani (2012) praised the company's hackathon initiative,

> Hackathons are a chance for engineers, and anyone else in the company, to transform the spark of an idea into a working prototype and get other people excited about its potential. We're a culture of builders, and hackathons are our time to take any idea—big or small, sane or crazy—and build it into something real for people to react to. Instead of worrying if their idea will scale for more than 900 million people, people are able to focus on getting their basic project up and running so the broader team can quickly iterate to make it better.
>
> *(Keyani, 2012, n.p.)*

Other tech industry leaders, such as 3M, Google, and Microsoft, have also embraced a similar culture, providing dedicated spaces for experimental projects. Yet, as with any given culture, the ethos of a tradition (old or new) should be examined from its social, political, and economic significance. While making is normalized in industry, we must acknowledge issues of social class relevant to the Maker Movement as they are becoming more central to the shaping of policies involving work and training. In the socioeconomic context, the maker culture is critiqued by some as a political philosophy and social movement, as Mike Rose (2014) noted,

> By and large, the Makers Movement is a middle-class movement. Working-class folk have not had the luxury of discovering making and tinkering; they've been doing it all their lives to survive—and creating exchange networks to facilitate it. Somebody across the street or down the road is a mechanic, or is wise about home remedies, or does tile work, and you can swap your own skills and services for that expertise.
>
> *(n.p.)*

Apprehensive about its social implications, I resound Joyce Locke Carter's (2016) proclamation that to make is to assume agency. Making disrupts conventions about teaching and learning, producing and consuming, and governing and subordinating. The maker culture, as noted by Andrew Schrock (2014), is focused on a flexible openness that supports its members as they move from "peripheral participants" to potentially "longstanding members engaged in ongoing projects" (p. 17). And because of its long-established culture of sharing the Maker Movement intersects deeply with the open-source movement. Such an alignment, especially in the education context, revolutionizes the way we create and disseminate knowledge. The open educational resources (OER) and open-access publishing movements are evidence of this revolution.

Moreover, forward-thinking policy makers have come forth and recognized this revolutionary culture. The Obama White House (n.d.) had openly acknowledged and promoted the values of the Maker Movement in education through the "Nation of Makers" initiative. It hosted its first White House Maker Faire[1] in 2014, making Maker Education a household name among teachers. Many school districts are now funded to launch their own fabrication program, receiving major support from governmental branches such as the US Office of Educational Technology (n.d.) and the National Science Foundation (2015), as well as corporate donors.

In support of the Obama Administration's effort in promoting Maker Education, more than 200 universities and colleges had signed a joint letter to the president in June 2014 articulating the significance of the Maker Movement in American education (Executive Office of the President, 2014). These universities pledged to focus on cross-disciplinary collaboration and documenting innovation processes so products from the makerspaces can remain open access and open source. Increasingly, university makerspaces are positioning themselves to be an interdisciplinary hub. In "A Review of University Maker Spaces," Thomas Barrett and colleagues (2015) compared 40 university based makerspaces and found an overwhelmingly similar narrative among them that these spaces "provide a central location for many campuses trying to encourage multidisciplinary activity" (p. 14). This phenomenon provides a unique opportunity for the field of technical communication to participate in the movement and establish fruitful connections with disciplines that would enrich both the field and our students.

Toward Productive Disruption in Technical Communication

Just as compositionists are reimagining academic and professional writing instruction by having students compose multimodally—and have yielded meaningful outcomes—we should rethink technical communication pedagogy to focus less on genres (memos, feasibility reports, instruction manuals, user guides, research reports, etc.). We should, as Miles Kimball (2017) proposed, help students to learn "to use technologies of communication to bring about practical change" (p. 350). Both design thinking making are ideally situated to help do that.

Design thinking exposes students to problem-based learning and hands-on problem solving. Its attributes cultivate a user-centered mindset and its methodology helps students exercise an iterative design for continuous improvement. This design-centric approach to technical communication pedagogy can help revitalize our curriculum by paying attention to both process and product, creating a more holistic development of technical communicators. It will also help technical communication programs to stay connected with industry practices by integrating design thinking with courses across different areas, including technical editing, content strategy, publication management, information architecture, user experience research, interaction design, documentation, and others.

Powered by design thinking, making foregrounds the materiality of problem solving and helps technical communication students understand the object dimension of problems. Through prototyping, students learned to identify the affordances and limitations in their different tools and materials, gaining an additional competency that is desirable in today's entrepreneurial culture. Making also fosters an environment for learning where individuals share ideas, experiment different approaches, and work on projects together. For technical communication, making invites learners to combine resources to tackle complex communicative issues. Such tendency is deemed favorable by public and private sectors today where collective intelligence (Levy, 2000) is considered valuable in social capital. Thus, to integrate such learning with technical communication pedagogy is to prepare students for their professional futures, where collaboration and cross-functional teams are already commonplace.

Together, making and design thinking spur innovations that respond to sociotechnical issues. Students and practitioners alike can take on wicked problems in technical communication (and beyond) through design challenges, which employ design thinking principles and maker culture. As user advocates, technical communicators should embrace social innovation as part of their professional responsibilities. As I have discussed in a previous chapter and in this conclusion, such productive disruption is necessary for today's social conditions. Technical communicators should "disrupt the future" (Jones, Moore, & Walton, 2016) using their socially and human-centered design skills and the available tools for meaningful changemaking. Our profession is fertile for such leadership and direction.

Summary and Future Directions

The future of technical communication should see its constituents as leaders in radical innovation for solving complex problems. In this book, I have taken up the question, "What can design thinking offer technical communication?" and I followed the social advocacy trajectory in current technical communication scholarship as a way to address this question. I have introduced and described design thinking as a mindset and a methodology for social innovation, which led to the focus on making as a pedagogical strategy for teaching user-centered design as well as a collaborative approach to solving problems. The pages between these covers have provided ethnographic findings, expert insights, pedagogical results, and self-study discovery that showcase the values of making and design thinking for technical communication. Adopting design thinking and making in our pedagogy and professional practice can help invigorate and modernize our approaches to technical communication but more importantly they keep us focused on the users.

In closing, I recommend the following research questions to further investigate the affordances (and limitations) of design thinking and making:

- What research method/ologies are most useful for studying design thinking and making processes? What should be our object(s) of study?
- How can the design thinking phases be updated to better facilitate technical communication needs or development?
- How can rhetorical thinking be integrated with design thinking and making?
- How can design thinking and making be more inclusive? How can we better include multiply-minoritized voices in design thinking and making?
- What innovative approaches might we take to promote (radical) collaboration?
- How can social innovation and social justice advocacy be emphasized in our profession?

These questions can serve as entry points toward productive research and exploration. There are certainly more questions pertaining to specific interest areas. I look forward to reading more findings and discussions from our field in the near future.

Learning Activity: Make, Disrupt, Innovate a Pedagogy of Advocacy

Here is one for instructors, faculty, and program administrators. Julie Schell (2018), Executive Director of Executive Education in the School of Design and Creative Technologies in the College of Fine Arts at The University of Texas at Austin, has made an important observation about design thinking pedagogy:

> Design thinking educators realize that one cannot effectively teach a novice to use human-centered design to solve vexing work or social problems using an accelerated pedagogical model. Teaching and learning design thinking in a way that results in impact requires slowing down the learning, taking time to unfold the layers of what it means to be human-centered and pay attention to the innate dignity of human beings.
>
> *(n.p.)*

Indeed, we are in many ways working in silos when teaching design thinking, and that makes it challenging to create real impact through our pedagogy within the conventional curricular structures. If you had the opportunity to reimagine academic programs and higher education, what would your responses to the following questions be?

Making to Advocate

- What management systems, collaboration technologies, and networking platforms can we build to facilitate community engagement with advocacy work?
- How can community partners provide leadership in our making of these technologies?
- What resources would we need to mobilize such efforts?

Disrupting to Advocate

- What ideologies, traditions, and practices do we need to disrupt in order to actualize social advocacy?
- What do these disruptions look like?
- Who may be affected by these disruptions?

Innovating to Advocate

- How can we foster innovative partnerships with our colleagues from disciplines such as architecture, computer science, fine arts, and human factors to create inter/cross-disciplinary programs that enable advocacy works?

- How can we make time for innovation?
- How can we keep our focus on social justice while innovating for advocacy?

Note

1. Since the change of administration after the 2016 presidential election, the White House has not provided official support to the Maker Movement nor hosted any Maker Faire. The Nation of Maker project can be found in the Obama Administration's web archives.

References

Barrett, T., Pizzico, M., Levy, B., & Nagel, R. (2015). A review of university maker spaces. In *Proceedings of 122nd ASEE annual conference and exposition* (pp. 1–17). Washington, DC: American Society for Engineering Education.

Bower, J., & Christensen, C. (1995). Disruptive technologies: Catching the wave. *Harvard Business Review*. Retrieved from https://hbr.org/1995/01/disruptive-technologies-catching-the-wave

Carter, J. L. (2016). Making, disrupting, innovating. *College Composition and Communication, 68*(2), 378–408.

Executive Office of the President. (2014). Building a nation of makers: Universities and colleges pledge to expand opportunities to make. Retrieved from www.whitehouse.gov/sites/default/files/microsites/ostp/building_a_nation_of_makers.pdf

Fogg, R. (2018). Catching disruption in the act: 3 problems innovation will solve in healthcare delivery. Christensen Institute. Retrieved from www.christenseninstitute.org/blog/three-problems-disruptive-innovation-solve-healthcare-delivery/

Holman, W. (2015). Makerspaces: Towards a new civic infrastructure. *Places Journal*. Retrieved from https://placesjournal.org/article/makerspace-towards-a-new-civic-infrastructure/?gclid=CMmk3vvxhtECFUO4wAodQYAOTg

Jones, N. N., Moore, K. R., & Walton, R. (2016). Disrupting the past to disrupt the future: An antenarrative of technical communication. *Technical Communication Quarterly, 25*(4), 211–229.

Keyani, P. (2012). Stay focused and keep hacking. *Facebook Engineering*. Retrieved from www.facebook.com/notes/facebook-engineering/stay-focused-and-keep-hacking/10150842676418920/

Kimball, M. (2017). The golden age of technical communication. *Journal of Technical Writing and Communication, 47*(3), 330–358.

Levy, P. (2000). *Collective intelligence: Mankind's emerging world of cyberspace*. New York, NY: Perseus Book Group.

National Science Foundation. (2015). Dear colleague letter: Enabling the future of making to catalyze new approaches in STEM learning and innovation. Retrieved from www.nsf.gov/pubs/2015/nsf15086/nsf15086.jsp

Obama White House. (n.d.). Nation of makers. Retrieved from www.whitehouse.gov/nation-of-makers

Rose, M. (2014). The maker movement: Tinkering with the idea that college is for everyone. *Truthdig*. Retrieved from www.truthdig.com/articles/the-maker-movement-tinkering-with-the-idea-that-college-is-for-everyone/

Schell, J. (2018). Design thinking has a pedagogy problem. . . and a way forward. *Journal of Design and Creative Technologies*. Retrieved from https://designcreativetech.utexas.edu/design-thinking-has-pedagogy-problem-way-forward

Schrock, A. R. (2014). Education in disguise: Culture of a hacker and maker space. *Inter-Actions: UCLA Journal of Education and Information Studies*, *10*(1), 1–25. Retrieved from https://escholarship.org/uc/item/0js1n1qg

Shelton, C. (2019). Shifting out of neutral: Centering difference, bias, and social justice in a business communication course. *Technical Communication Quarterly*. Online first version. Retrieved from www.tandfonline.com/doi/full/10.1080/10572252.2019.1640287

Smith, A. (2014). Technology networks for socially useful production. *Journal of Peer Production*, *5*. Retrieved from http://peerproduction.net/issues/issue-5-shared-machine-shops/peer-reviewed-articles/technology-networks-for-socially-useful-production/

US Office of Educational Technology. (n.d.). Makerspaces. Retrieved from http://tech.ed.gov/stories/makerspaces/#

APPENDIX A

Radical Collaboration Survey Questionnaire

Name: Your Role:

Please use a word or a phrase to describe your overall experience as a member of the WRC this semester.
In what ways was *invitation* achieved and/or not achieved in your experience with the WRC this semester?   To what degree do you think *invitation* was actualized this semester: 0-1-2-3-4-5
In what ways was *sharing* achieved and/or not achieved in your experience with the WRC this semester?   To what degree do you think *sharing* was actualized this semester: 0-1-2-3-4-5
In what ways was *collaboration* achieved and/or not achieved in your experience with the WRC this semester?   To what degree do you think *collaboration* was actualized this semester: 0-1-2-3-4-5

In what ways was *radical imagination* achieved and/or not achieved in your experience with the WRC this semester?
To what degree do you think *radical imagination* was actualized this semester: 0-1-2-3-4-5
In what ways was *suspension* achieved and/or not achieved in your experience with the WRC this semester?
To what degree do you think *suspension* was actualized this semester: 0-1-2-3-4-5
In what ways was *exposure* achieved and/or not achieved in your experience with the WRC this semester?
To what degree do you think *exposure* was actualized this semester: 0-1-2-3-4-5

The dimensions/features of Radical Collaboration:

- Invite and welcome perspectives that span theoretical, personal, and professional boundaries;
- Expose participants to the complexities of problems regardless of experience;
- Invite radical imagination to what learning in academia can mean and be;
- Share leadership, research, and teaching roles;
- Suspend beliefs about knowledge boundaries; Suspend judgement of people and ideas; Suspend closure; and Sustain openness. You need to
- Resist hierarchical structures; Invite and welcome perspectives across institutional boundaries; and Value team learning

APPENDIX B
Design Thinking Methods and Exercises

Overview: This appendix contains some signature design exercises and research methods employed by design thinkers and makers. While it is organized by the design thinking process, it does not suggest these methods and exercises are to be applied in a linear fashion.

Empathy	
Bodystorming	See UX Advocacy Methods in Chapter 3, p. 74
Contextual inquiry	See Pedagogical Exercises in Chapter 4, p. 95
Journalistic questioning	See UX Advocacy Methods in Chapter 3, p. 72
Journey mapping	See UX Advocacy Methods in Chapter 3, p. 74
Photovoice	See UX Advocacy Methods in Chapter 3, p. 73

Definition	
Asking "How might we?" (HMW)	HMW questions are questions that spark ideas during open-ended brainstorming sessions. In order to perform this exercise, you should have your POV statement ready. For example, say your POV is this: "First-year employees need to learn about company cultures in order to participate in company activities meaningfully" You may ask HMW questions like: • How might we make company cultures learnable? • How might we inspire new employees to participate in company activities? • How might we create meaningful activities? The goal of this exercise is to create actionable (hence "how" questions) statements that can guide the ideation phase later.

(*Continued*)

Empathy mapping	See UX Advocacy Methods in Chapter 3, p. 73
Point-of-view (POV) statement	See Pedagogical Exercises in Chapter 4, p. 96

Ideation

Affinity mapping or affinity diagramming	Affinity mapping is borrowed from UX design to help teams organize related ideas into distinct clusters or categories. This is typically done with post-it notes. First, the team leader presents all available ideas on a wall. Looking at these ideas, the entire team helps to create top-level categories that are then split into subcategories that could house all of the ideas. All of these categories/subcategories should be clearly labeled (and better if color coded). The last step is to organize all ideas into these subcategories. It is best to put categories that are close to each other (in nature, execution, etc.) so team members can visually identify the relationships (hence "affinity"). When done organizing the ideas into categories, team members will take turns to present the ideas within a category and summarize them in simple narratives.
Dot voting	Depending on the materials used for this activity, it is sometimes called post-it voting. Using a given number of dot stickers or post-its, each member casts their votes to the corresponding ideas they like best. This simple exercise is a quick way for team members to choose their favorite ideas and shortlist options.
Four-category mapping	This method involves dividing generated ideas into relative abstract categories: the rational, the delightful, the darling, and the long shot. Similar to affinity mapping, team members should first have a view of all available ideas, then they assign these ideas into either of the four categories. The purpose of these categories is to create a set of *doable* ideas and those that may require more resources to accomplish, and may well be saved for the future. For some design teams, this exercise allows them to combine ideas to create hybrid yet achievable solutions, e.g., both rational and darling.
Radical imagination	See Pedagogical Exercises in Chapter 4, p. 96
Storyboarding	Storyboard is an extensive ideation method borrowed from motion picture/film/animation practices. It requires team members to sketch a visual representation of their imagined user journey using different personas and user stories. A storyboard has a beginning and an end, and usually follows the traditional story arc (starting, rising action, climax, anti-climax, resolution). The combination of words and images help create a visible user scenario that can guide designers in understanding user experience (their emotions, pain points, etc.).

Prototyping	
3D modeling	This is a technique for producing 3D digital representations of any object design using graphic software like TinkerCAD and Vectary. Some 3D printers come with their associated modeling software. The purpose of performing 3D modeling is to evaluate a design before it is produced into the physical world using tools like 3D printers or CNC milling. It saves manufacturing costs.
3D printing	Also known as additive manufacturing, 3D printing is the process of outputting 3D models using solid materials like carbon fibers. To print in 3D, the printing application takes a 3D model file and slices/divides it into hundreds or thousands of horizontal layers, and feeds the slices to a printer to print them layer by layer.
Computer-numerically controlled (CNC) milling	CNC milling is the process of cutting or drilling materials—wood or metal—using a rotating cylindrical cutter controlled by a computer. Like 3D printing, the milling machine is informed by a 3D model file that determines which point on the material the cylindrical cutter should go, in what angle, and along which axes (X, Y, Z) it should move.
Laser/waterjet cutting	Both of these cutting or engraving methods are popular in the prototyping process to quickly mark or cut thick or tough surfaces (like metal or glass). Like CNC milling, the laser beam or water jet nozzles are controlled by digital files.
Paper prototype	This low-cost prototyping method uses papers and pens to create mock-up interfaces that can be used for quick user testing. You just need to sketch a representative image of the interface design, using stick features, boxes, and scribbles to represent people, buttons, and texts on a piece of paper. It saves time by eliminating professional coding and graphic design in the early stage of design.
Wireframing	Wireframing is the next step from paper prototypes. It is usually done using a graphic software like PowerPoint or InDesign to create mid-fidelity visual representations of a user interface. Applications like InVision and Axure allow designers to create clickable wireframes. The goal of wireframing is to present a clean, bare-bone structure of a layout that can give users a sense of the overall composition without the rich content like images, colors, and copy (texts).

Testing	
Card sorting	This is a method used by designers to evaluate the information architecture of a website or application. There are open and closed card sorting methods. In open card sorting, participants (users) are asked to organize various topics from a potential website into groups *they define*, like home,

(*Continued*)

	resources, people, and contact. Doing so allows designers to understand what labels work best to group different topics. In closed card sorting, participants will sort topics from your content into predefined categories (you give them the labels).

This exercise helps designers create navigation or interaction logic that matches the user's expectations.

Heuristic evaluation	Heuristic evaluation is a straightforward and practical usability testing method that uses Jakob Nielsen and Rolf Molich's 10 usability heuristics:

1. Visibility of system status (user knows what's going on)
2. Match between system and the real world (follows logical conventions)
3. User control and freedom (easy to perform actions and recover from errors)
4. Consistency (standardize words, actions, or feedback)
5. Error prevention (minimize potential user mistakes)
6. Recognition rather than recall (Minimize memory load)
7. Flexibility and efficiency (has "accelerators" that advanced users can apply to speed up actions)
8. Aesthetic (minimalist design)
9. Help users recognize, diagnose, and recover from mistake (this is self-explanatory)
10. Help and documentation (proper guides are available when needed)

You may invite a group of 3–5 potential users to participate in a heuristic evaluation session of your design, and ask them to rate the above 10 principles from no issue to severe issue based on their interaction/testing with your design.

Think-aloud protocol	A very popular method used by UX researchers, the think-aloud protocol requires a setup where the user/participant can experiment with your design via a semi-guided manner. First, you generate a set of scenarios and tasks that can be done using your design. Then, you provide minimal instructions to the user and ask them to attempt completing the tasks. When doing so, ask them to speak out their thoughts (hence "think-aloud") so you can "see" what their thought process or problem-solving approach is, as well as their feelings. This method is usually supplemented with a post-session interview where you may ask the user what they thought was the easiest or most difficult task(s) to perform and why.

APPENDIX C
Annotated Bibliography

Overview: Because the topics of design thinking and making are relatively fresh to technical communication, I include here annotated summaries of 15 scholarly publications mentioned in the book. I consider these selected publications essential for understanding the theory and practice of design thinking and making. May these be useful for instructors who are (re)designing their courses, researchers who are seeking peer-reviewed sources to support their studies, and graduate students who are putting together their exams or dissertation reading lists.

Bay, J., Johnson–Sheehan, R., & Cook, D. (2018). Design thinking via experiential learning: Thinking like an entrepreneur in technical communication courses. *Programmatic Perspectives, 10*(1), 172–200. Retrieved from https://cptsc.org/wp-content/uploads/2018/06/vol10.1.pdf

Using the technical communication service course as a backdrop, Bay, Johnson-Sheehan, and Cook explored an "entrepreneurship pedagogy" through the lens of design thinking. The article includes an expanded description for each phase of design thinking—empathy, definition, ideation, prototyping, and testing. Bay et al. connected design thinking with experiential learning. Combined with entrepreneurial thinking, this pedagogical approach can invigorate technical communication programs by creating (and sustaining) working relationships with industry. Bay et al. provided some examples of such collaboration between the Purdue program and entrepreneurship hubs in the city of Lafayette.

Breaux, C. (2017). Why making? *Computers and Composition, 44,* 27–35.

Breaux synthesized key historic movements and influences that help establish the lineage of maker culture as we know it today. Through assemblage/
remediation, craft, and hacking theories, Breaux enacted a strong intersection between making and a subfield of writing studies—computers and
writing. Breaux then drew three major implications of making for writing:
1) making expands the constructions of literacy, 2) making shines a spotlight on underrepresented groups in technology studies, and 3) making
supports democratization of technology (production and consumption).
For these reasons, Breaux argued that computers and writing teacher-
scholars should consider themselves as makers and justify the importance
of making.

Brown, J., & Rivers, N. (2013). Composing the carpenter's workshop.
O-Zone: A Journal of Object-Oriented Studies, 1(1), 27–36.

Given the field's focus on rhetorical ecology, objects, and agency, Brown
and Rivers contended that rhetoric and composition can be hospitable to
projects concerning object-oriented production. By the way of Ian Bogost's
"philosophical carpentry," Brown and Rivers enacted rhetorical carpentry
as an attunement to multimodal composition and object-oriented rhetoric. This ecological and ontological approach to writing instruction can
introduce students "to a multiplicity of composing skills" by rendering
experiences of both humans and nonhumans, moving beyond just the traditionally human-centered rhetorical situation.

Carter, J. L. (2016). Making, disrupting, innovating. *College Composition and Communication, 68*(2), 378–408. Retrieved from https://secure.
ncte.org/library/NCTEFiles/Resources/Journals/CCC/0682-dec2016/
CCC0682Address.pdf

This print article was the fifth iteration of the Chair's Address given by
Carter, 2016 CCCC Chair, at the national convention in Houston. Carter
urged writing scholars and instructors to consider themselves as makers
and innovators who have always-already disrupted so-called norms. Carter
provided examples of in-house innovations—those produced by "writing"
scholars and have made a difference in teaching and learning, publishing,
and even manufacturing. Carter equated writing to coding by demonstrating the similarities between argumentation schematics, linguistic diagram-
ming, poetry, and algorithm programming. Carter argued that making,
disrupting, and innovating should not be seen as anomalies in writing
studies.

Hailey, D., Cox, M., & Loader, E. (2010). Relationship between innovation and professional communication in the "creative" economy. *Journal of Technical Writing and Communication, 40*(2), 125–141.

To examine the impact of creativity on professional communication careers, Hailey, Cox, and Loader, together with a group of students, used a variation of the Delphi method to evaluate 45 selected job titles, assigning them into "more" or "less" creative categories. The study highlighted the offshoring (outsourcing) phenomenon in a "creative" economy. Results showed that creativity does not equate innovation, and that individuals who can demonstrate knowledge of innovation processes are more valued by industry than those who were mere creative. The authors suggested that technical communicators who can consistently identify and solve corporate problems will be more valuable. To this end, the authors provided an 8-step "innovation process for writers" that focuses on problem definition, developing solutions, and communicating them to stakeholders.

Johnson-Eilola, J., & Selber, S. (2013). *Solving problems in technical communication.* Chicago, IL: University of Chicago Press.

This edited collection consists of 19 chapters, organized into a four-phase adaptive heuristic that spans from broad to specific technical communication contexts and applications. Johnson-Eilola and Selber's introduction to this collection characterizes technical communication as a problem-solving activity, and states that students and professionals should be equipped to approach ill-structured problems. Each chapter provides a rationale and context for the chapter subject matter, a literature review to situate the subject within existing discussions, a research-based heuristic that serves as a framework for practice, an extended example to provide an illustration of the framework at work, and a conclusion with discussion questions. Essential chapters include Bill Hart-Davidson on work patterns of technical communication, Jason Swarts on technical communication work tools, and Rebecca Burnett et al. on collaboration.

Jones, N. N., Moore, K. R., & Walton, R. (2016). Disrupting the past to disrupt the future: An antenarrative of technical communication. *Technical Communication Quarterly, 25*(4), 211–229.

Recognizing that the field of technical communication has mostly focused o "objective, apolitical, acultural practices, theories, and pedagogies," Jones, Moore, and Walton were motivated to present an "antenarrative" for the field that embraces social justice and inclusivity as a core tenet of technical communication. Jones et al. examined the dominant narrative in the field's published literature and then presented a collection of nondominant threads that "unravel" the dominant narratives. These threads come from feminism and gender studies, race and ethnicity, international/intercultural professional communication, community and public engagement, user advocacy, and disability and accessibility. The antenarrative threads invite reinterpretation of the past and open room for a more inclusive future. The authors offered a heuristic approach—3P: positionality, privilege, and power—to inform inclusive technical communication practice and scholarship.

Knievel, M. (2006). Technology artifacts, instrumentalism, and the *Humanist Manifestos*: Toward an integrated humanistic profile for technical communication. *Journal of Business and Technical Communication, 20*(1), 65–86.

Knievel interrogated the humanistic tradition and debates in technical communication to identify their implications for the role of technology in the field's discourse. Knievel noted that traditional treatment of technology as instrumental, not rhetorical, within the field's humanistic profile raises questions about the legitimacy of the field's humanistic status. Through the *Humanist Manifestos*, Knievel encouraged scholars to reconceptualize the relationship between technology and humanity by acknowledging the "full range of technology—instrumental, rhetorical, systemic, or substantive." Knievel offered three possible ways to spread the notion of humanistic technology: 1) publish in English studies and humanistic journals, 2) collaborate with business and industry, and 3) bring technology courses into English departments.

Kostelnick, C. (1989). Process paradigm in design and composition: Affinities and directions. *College Composition and Communication, 40*(3), 267–281.

Kostelnick recognized some overlaps between traditional composition and the design process, and was set out to examine the affinities between the two movements. The common tenets shared by these movements include writing and designing as acts of discovery, writing and designing as recursive invention, the consciousness of experienced writers and designers of their own processes, and the role of audience analysis in defining writing and design problems. Kostelnick argued that composition scholars can learn from the methods crisis in design, and adopt a pluralistic approach to process pedagogy. Kostelnick urged composition studies to develop processes similar to design pedagogies and theories that "reconcile the writing process paradigm with real world text production."

Leverenz, C. (2014). Design thinking and the wicked problem of teaching writing. *Computers and Composition, 33*, 1–12.

Concerned with the increasing demand of multimodal composing in the classroom and workplace, Leverenz considered how writing courses can be reimagined as opportunities for design thinking. By tracing the argument for design thinking in composition studies to Richard Buchanan (1992) and Richard Marback (2009), Leverenz argued that a human-centered approach to designing innovative solutions in response to wicked problems can better students for their future of writing. Leverenz suggested that writing assignments can be made like design briefs to represent real design problems. To simulate workplace practice, Leverenz also recommended team-based writing and letting students experiment with new ways of thinking through prototyping. Leverenz offered a sample assignment sequence she used in junior-level class to show how the elements of design thinking could be applied to writing.

Marback, R. (2009). Embracing wicked problems: The turn to design in composition studies. *College Composition and Communication, 61*, 397–419.

Marback agreed that the concept of design appeals to composition studies but it has yet to be developed enough to benefit composition pedagogy. Like Charles Kostelnick (1989), Marback believed that composition studies can benefit from a flexible design paradigm, with focus on the idea of design tasks as wicked problems. By attending to wicked problems as technical as well as rhetorical, Marback sought to "return" design to composition studies. Marback critiqued the efforts by the New London Group (1996), Diana George (2002), and Mary E. Hocks (2003), arguing that design thinking is not about the extension of print media to other meaning-making platforms, but rather embracing the wickedness of design problems. Marback presented a sample assignment to demonstrate this level of engagement in a composition classroom.

Purdy, J. (2014). What can design thinking offer writing studies? *College Composition and Communication, 65*(4), 612–641.

Purdy followed Richard Marback's (2009) call to turn to design in composition studies and explored why design is invoked in five writing studies and computers and composition journals (from their inception to 2011). Purdy's study yielded five categories of design use: 1) design as a synonym for plan/structure, 2) design as a conceptualization of multimodal composing, 3) design as a recognition of digital/multimedia compositions, 4) design as attending the material conditions of composing, and 5) design as a discussion of the discipline of design studies. Purdy then synthesized the ways design is approached in writing studies pedagogy, and attempted to align the prevalent steps of design thinking with the writing process. To demonstrate the application of design thinking in writing studies, Purdy shared an example from the Colorado State University Writing Project, highlighting the key phases of design thinking in executing the multimodal project. Purdy contended that design thinking can cast focus beyond print composition, help form collaborative partnerships, and (re)orient writing as productive work in the world.

Sheridan, D. (2010). Fabricating consent: Three-dimensional objects as rhetorical compositions. *Computers and Composition, 27*(4), 249–265.

Through the lens of material rhetoric, Sheridan explored the implications of three-dimensional fabrication of products for composition and rhetoric. Drawing from early discussions of visual and multimodal rhetoric, Sheridan submitted four reasons for integrating 3D rhetoric into writing pedagogy—it's possible (access to technologies of production), it's powerful (more effective communication), it's valued (deemed important across personal, professional, and public spheres), and it's ours (it resides in the domain of rhetoric). Sheridan compiled literature that supported these reasonings and presented three brief cases of citizen fabrication that explored arguments from the perspectives of infrastructural accessibility, rhetorical effectiveness, cultural status, and (de)specialization in 3D rhetoric. Sheridan posited 3D fabrication as transformative to cultures and politics, and can help composition and rhetoric achieve the goals of equality and social justice.

Shipka, J. (2005). A multimodal task-based framework for composing. *College Composition and Communication, 57*(2), 277–306.

In this study, Shipka followed two students in their multimodal composing journey to reconceptualize production, delivery, and reception in the composition classroom. Through theories of goal-oriented activity, Shipka examined how the students worked to negotiate the complex communicative tasks they undertook in a class. Shipka reported that these students, when called upon to set their own goals and strategies to accomplish those goals, they can 1) demonstrate awareness of media affordances; 2) successfully produce, represent, distribute, and deliver their work; and 3) become better equipped to negotiate the range of communicative contexts they found themselves in. Based on these accounts, Shipka argued for a multimodal, task-based framework for composing that asks students to attend to: a) the product(s) they will formulate in response to a given task; b) the operations, processes, or methodologies that will (or could) be employed in generating that product; c) the resources, materials, and technologies that will (or could) be employed in generating that product; and d) the specific conditions in, under, or with which the final product will be experienced.

Wible, S. (2020). Using design thinking to teach creative problem solving in writing courses. *College Composition and Communication, 71*(3), 399–425.

Motivated by the increased interest in helping students foster creativity through writing courses, Wible integrated design thinking methods as a creative problem-solving process. Through genre theory, Wible examined what specific genres were written and discussed by students in their design thinking-oriented writing course. In this article, Wible focused on the problem definition, invention, and prototyping phases in the design thinking process. Wible noted that students used design thinking genres such as user empathy maps and point-of-view statements to formulate problem definitions. Students collaboratively brainstormed solutions using how-might-we questions, selected creative ideas through voting, composed multimodal prototypes using storyboards, and tested them with actual users. Finally, students gave an oral presentation via either of two narratives—an innovation story or a learning story. Wible concluded that design thinking can help students develop creativity and problem-solving skills through audience and purpose-focused genres that are different from typical patterns of inquiry and argumentation in traditional writing courses.

INDEX

Note: Page numbers in *italic* indicate a figure and page numbers in **bold** indicate a table on the corresponding page.

academic rank **107**
academic settings 29, 32; *see also* higher
    learning
access 38–42, 44–46
actualization *see* degree of actualization
advocacy 61–62, 90; user advocacy
    methods through design thinking
    72–74; *see also* social advocacy
Anderson Student Innovation Labs *see*
    University of Minnesota–Twin Cities
    (UMN)
assignments 81–85, 87–90; correlating
    goals and **84**; descriptions and weight in
    percentage **85**
assignment sequence 81–85, *83*, 87–88
Association for Teachers of Technical
    Writing (ATTW) 16, 100
audit 116–117
auditing worksheet 116, **117**
autoethnography *see* collaborative
    autoethnography

bias toward action 102
bodystorming 10, 74, 133

Case Western Reserve University: Sears
    Think[box] 44–50
Castillo, Vivianne 60, *60*

change and changemaking 15, 67, 80,
    121–123; advocating for 90; technical
    communication as 57–61
clarity *see* craft clarity
collaboration 109–110; and design
    thinking attributes 101–103; in
    technical communication 99–101; *see
    also* collaborative autoethnography;
    radical collaboration
collaborative autoethnography (CAE) 18,
    105–107, *106*, 112–113, 118
contextual inquiry 10, 79, 87, 95–96
Council for Programs in Technical and
    Scientific Communication (CPTSC)
    100
craft clarity 102

datapoints 86–90
definition 10, 20, 87–88
degree of actualization 106, 112, *112*, **112**
design *see* human-centered design; medical
    design; participatory design; user-centered
    design; user experience (UX) design
design challenge 18, 82–84, 86–91, **91**,
    93–95, 97–98
design making 12–15; and technical
    communication pedagogy 17–18; *see
    also* making

design thinking: 121–122; basic model adapted from the Stanford model *10*; a brief historical sketch 6–8; data points 86–90; designing design thinking in technical communication pedagogy 81–86; the design (thinking) turn 1–5; future directions 127; key developments of *6*; learning activities 19–21, 98, 128–129; making as 50–52; methodology for socially just and ethical innovation *71*; methods and exercises 133–136; mindset and methodology 8–11; pedagogical exercises 95–97; as productive disruption 122–123; a rationale for 81; student projects 90–93; student responses 93–95; and technical communication pedagogy 17–18; user- and human-centered design 11–12; and wicked problems 15–17
design thinking attributes 101–103
design thinking process 11, 82–84, *83*, **84**
digital fabrication labs (fablabs) 31; *see also* makerspaces
disciplinary contexts **114**
discussions 76, 116–117
disruption: design thinking as productive disruption 122–123; future directions 127; learning activity 128–129; making as a disruptive culture 124–126; toward productive disruption in technical communication 126–127

Ecological Momentary Assessment Robot (EMAR) 63
empathy 9–10, 20, 87; and Google Glass 64–65
empathy mapping 73–74, *73*
ethics 61–63, *71*
experimentation 30, 51, 79–80, 102
exposure 102, 111–113

fablabs *see* digital fabrication labs
Fuller, Buckminster 6–7

Georgia Institute of Technology: Invention Studio 41–44, 48–50
goals 20, 80–81, **84**
Google Glass 64–66, 103, 122–123

hackerspaces 31–32; *see also* makerspaces
Higher Education Makerspaces Initiative (HEMI) 32, 36

higher learning: Anderson Student Innovation Labs at UMN 37–41; the Invention Studio at Georgia Institute of Technology 41–44; learning activity 52–54; makerspaces in 36–37, 48–50; opportunities for technical communication 50–52; the Sears Think[box] at Case Western Reserve University 44–48
historical sketch 6–8
homophobia 60, *60*
human-centered design 7, 11–12, 63, 101, 127, 128
human experience: empathy map 73–74, *73*
human values 101

ideas 10–12, 20–21, 124; and the Maker Movement 34–35, 47–49; and radical collaboration 102–103, 105, 110–111, 115–118; and social advocacy 80–81, 83–84, 88–90, 96–97
ideation 10, 20, 88–89
IDEO 7, 70
implementation 84, 117
inControl 64, 65
industrialist legacy 31–36
industry views 67–70
innovation 121 128; learning activity 128–129; *see also* social innovation
institutional contexts **114**
International Symposium on Academic Makerspaces (ISAM) 32, 44
Invention Studio *see* Georgia Institute of Technology
invitation 102, 107–108
IT: social innovation in 67–68
iterative processes 31, *106*

journalistic questioning 72
journey mapping 10, 74

leadership: and social innovation 70–72
learning activities: a(nother) design challenge 98; considering dilemmas in radical collaboration 118; design thinking orientation 19–21; facilitate a community workshop 75–76; make, disrupt, innovate a pedagogy of advocacy 128–129; transforming a classroom into a makerspace 52–54

Make: magazine 31–32, 35
Maker Education 14, 125
maker experience: at Anderson Labs
39–41; at Invention Studio 42–44; at
Think[box] 46–48
Maker Faires 31, 35–36, 129n1
Maker Movement 12; industrialist
legacy of 31–36; and the materiality
of technical communication and its
pedagogy 27–31; opportunities for
technical communication 50–52; *see
also* makerspaces
makerspaces 36–37, 48–50, *53*; Anderson
Student Innovation Lab at UMN
37–41; the Invention Studio at Georgia
Institute of Technology 41–44; learning
activity 52–54; the Sears Think[box]
at Case Western Reserve University
44–48
making 12–15: in academic settings
36–37; datapoints 86–90; as design
thinking 50–52; as a disruptive
culture 124–126; future directions
127; learning activities 19–21, 98,
128–129; pedagogical exercises
95–97; a rationale for 81; student
projects 90–93; student responses
93–95; and technical communication
pedagogy 17–18, 79–81; *see also*
design making; Maker Movement;
makerspaces
materiality 18, 27–31, 126
medical design, 68, 70
methodology, design thinking 8–11, 18,
70, 79, 94–95, 98
methods 18–19, 133–136
mindset, design thinking 8–11, 19, 70,
122–123

Online Writing Instruction (OWI) 16
open access 32, 44–45, 62, 124–125
open educational resources (OER) 125
open source 100, 125

parameters 87–88
participatory design 7, 61, 63, 65, 75, 101
pedagogy: datapoints 86–90; exercises
95–97; learning activity 98; student
projects 90–93; student responses
93–95; technical communication
pedagogy 15–18, 27–31, 79–86
photovoice 10, 73
plans 116–117

point-of-view (POV) statement 20, 76,
96, 133–134
post-audit discussion 116–117
pre-audit reflection 116
privilege 60, *60*, 64, 75
problem-solving 17–19, 50–51, 81–82
process 2–3
programmatic contexts **114**
prototyping 10–11, 20–21, 97, 117;
inControl 64; testable solutions 89

questions 21; and disruption 123, 127,
128; journalistic questioning 72; and
radical collaboration 105–107, 116,
117, 118; and social advocacy **85**, 98;
and social innovation 67, 75–76

racism 60, *60*, 62
radical collaboration 103–105; auditing
worksheet **117**; degree of actualization
of *112*, **112**; designing project
workflow 116–117; and design thinking
attributes 101–103; learning activity
118; methodology 105–106; model
for *113*; results 106–113; strategies for
cultivating **114**; supporting a culture
of 113–115; survey questionnaire
131–132; in technical communication
99–101
radical imagination 96–97, 103, 110

Sears Think[box] *see* Case Western
Reserve University
sharing 102, 108–109
show, not tell 101
social advocacy: datapoints 86–90; learning
activity 98; pedagogical exercises
95–97; student projects 90–93; student
responses 93–95
social change 58, 67, 80, 90; *see also*
change and change-making
social innovation: in academic UX services
69–70; a call to advocacy 61–62; in and
out of the classroom 70–72; examples
62–65; implications for technical
communication 65–66; in IT 67–68;
learning activity 75–76; in medical
design 68; technical communication
as changemaking 57–61; in technical
documentation 69; user advocacy
methods through design thinking 72–74
social justice 58–59, 62–64, 71–72, *71*,
74–75

Stanford d.school 7, 10, *10*, 82
student projects 50, 90–93, 95
student responses 93–95
survey questionnaire 131–132
suspension 102, 110–111

technical communication: and advocacy
    61–62; the Anderson Student
    Innovation Labs at UMN 37–41;
    case study 103–113; as changemaking
    57–61; collaboration in 99–101;
    designing design thinking in technical
    communication pedagogy 81–86;
    future directions 127; implications
    of social innovation for 65–66; the
    Invention Studio at Georgia Institute
    of Technology 41–44; learning
    activities 52–54, 75–76; and the
    Maker Movement 31–36; makerspaces
    in 36–50; making in technical
    communication pedagogy 79–81; the
    materiality of 27–31; opportunities
    for 50–52; productive disruption in
    126–127; the Sears Think[box] at
    Case Western Reserve University
    44–48; social innovation examples
    62–65; social innovations in and out
    of the classroom 70–72; user advocacy
    methods through design thinking

72–74; views from the industry
    67–70; what design thinking and
    making offer 17–18; wicked problems
    in technical communication pedagogy
    15–17
technical documentation: social innovation
    in 69
techshops 31; *see also* makerspaces
testable solutions 89
testing 11–12, 21, 117
Twitter *60*

University of Minnesota–Twin Cities
    (UMN): Anderson Student Innovation
    Labs 37–41, 48–50
user advocacy 61–62, 66, 70, 72–75
user-centered design 11–12, 19, 61
user experience (UX) design 5, 11–12, 29,
    60–64, *60*, 68–71; social innovation in
    69–70

website 103–104, *104*
white supremacy *60*
wicked problems 3–4, 7–8, 15–17, 59, 83
workflow 37; at Invention Studio 42;
    and radical collaboration 116–117; at
    Think[box] 45–46

xenophobia 2, 60, *60*

Printed in the United States
by Baker & Taylor Publisher Services